Reasoning Algebraically About Operations

Facilitator's Guide

A collaborative project by the staff and
participants of Teaching to the Big Ideas

Principal Investigators

Deborah Schifter

Virginia Bastable

Susan Jo Russell

And teacher collaborators

DALE SEYMOUR PUBLICATIONS

Pearson Learning Group

National Science Foundation

ExxonMobil

This work was supported by the National Science Foundation under Grant Nos. ESI-9254393 (awarded to EDC), ESI-9731064 (awarded to EDC), ESI-0095450 (awarded to TERC), and ESI-0242609 (awarded to EDC). Any opinions, findings, conclusions, or recommendations expressed here are those of the authors and do not necessarily reflect the views of the National Science Foundation.

Additional support was provided by a grant from the ExxonMobil Foundation.

Art & Design: Evelyn Bauer, Kamau DeSilva
Editorial: Margie Richmond, Jennifer Chintala, Jennifer Serra
Production/Manufacturing: Nathan Kinney
Marketing: Kimberly Doster

ISBN-13: 978-1-4284-0516-5
ISBN-10: 1-4284-0516-X

Printed in the United States of America
2 3 4 5 6 7 8 9 10   11 10 09 08

Dale Seymour Publications
Pearson Learning Group

1-800-321-3106
www.pearsonschool.com

Teaching to the Big Ideas

The *Developing Mathematical Ideas* (DMI) series was conceived by Teaching to the Big Ideas, an NSF Teacher Enhancement Project. *Reasoning Algebraically About Operations* was developed as a collaborative project by the staff and teacher collaborators of the Teaching to the Big Ideas and Investigations Revisions Projects.

PROJECT DIRECTORS: Deborah Schifter (Education Development Center), Virginia Bastable (SummerMath for Teachers), and Susan Jo Russell (TERC).

CONSULTANTS: Elham Kazemi, Stephen Monk, Virginia Stimpson (University of Washington), Thomas Carpenter (University of Wisconsin at Madison), Herbert Clemens (Ohio State University), Mark Driscoll (Education Development Center), Benjamin Ford (Sonoma State University), Christopher Fraley (Lake Washington Public Schools), Megan Franke (University of California at Los Angeles), James Kaput (University of Massachusetts at Dartmouth), Jill Lester (Mount Holyoke College), James Lewis (University of Nebraska), Jean Moon (National Academy of Sciences), Loren Pitt (University of Virginia), Polly Wagner (Boston Public Schools), Erna Yackel (Purdue University Calumet).

TEACHER COLLABORATORS: Beth Alchek, Kim Beauregard, Barbara Bernard, Janelle Bradshaw, Nancy Buell, Rose Christiansen, Lisette Colon, Kim Cook, Fran Cooper, Maria D'Itria, Pat Erikson, Richard Feigenberg, Tom Fisher, Mike Flynn, Elaine Herzog, Kirsten Lee Howard, Liliana Klass, Melissa Lee, Jennifer Levitan, Solange Marsan, Kathe Millett, Florence Molyneaux, Elizabeth Monopoli, Robin Musser, Christine Norrman, Deborah Carey O'Brien, Mary Beth Cahill O'Connor, Anne Marie O'Reilly, Mark Paige, Crissy Pruitt, Margaret Riddle, Rebeka Eston Salemi, Karen Schweitzer, Lisa Seyferth, Shoshy Starr, Geri Smith, Susan Bush Smith, Liz Sweeney, Janice Szymaszek, Danielle Thorne, Karen Tobin, JoAnn Traushke, Ana Vaisenstein, Carol Walker, Yvonne Watson-Murrell, Michelle Woods, and Mary Wright, representing the public schools of Amherst, Boston, Brookline, Cambridge, Greenfield, Holyoke, Lincoln, Natick, Newton, Northampton, South Hadley, Southampton, Springfield, Sudbury, Westwood, and Williamsburg, MA, and the Smith College Campus School in Northampton, MA.

DVD DEVELOPMENT: David Smith (David Smith Productions).

FIELD-TEST SITES: Albuquerque Public Schools (New Mexico), Bedford County Public Schools (Virginia), Bismarck Public Schools (North Dakota), Boston Public Schools (Massachusetts), Buncombe County Public Schools (North Carolina), Durham Public Schools (North Carolina), Fayetteville Public Schools (Arkansas), Houston Independent School District (Texas), Lake Washington School District (Washington), Northampton Public Schools (Massachusetts), Seattle Public Schools (Washington), Stafford County Schools (Virginia), and Ventura Unified School District (California).

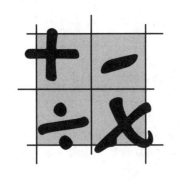

CONTENTS

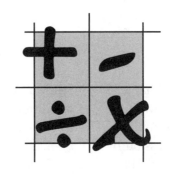

Orientation to the Materials

Number and Operations, Part 3: Reasoning Algebraically About Operations

Developing Mathematical Ideas (DMI) is a professional development curriculum designed to help teachers think through the major ideas of elementary and middle school mathematics and examine how students develop those ideas.

The Reasoning Algebraically About Operations (RAO) seminar provides the opportunity for teachers from Kindergarten through middle school to

■ Articulate generalizations *implicit* in students' work, as well as those that students state *explicitly*, as they notice properties of the operations and relationships among operations

■ Express these generalizations in natural language, in symbolic forms, with diagrams or drawings, and with physical models

■ Examine the connections among student thinking about these general claims, teachers' own thinking, and the formal statements known as the Laws of Arithmetic

■ Explore "representation-based proofs"— arguments based on diagrams and physical models—as well as proofs based on symbolic notation

The RAO seminar builds on the work of the two previous DMI Number and Operations seminars—Building a System of Tens (BST) and Making Meaning for Operations (MMO). To ensure that participants are in a position to gain the most from RAO, it is helpful if they have prior experience with the content of BST and MMO before enrolling in the RAO seminar. If the RAO seminar is to be offered to participants without this experience, then it is recommended that the number of seminar sessions be expanded from eight to twelve in order to begin the experience with the first four sessions of MMO and then continuing with RAO.

Components of the RAO Developing Mathematical Ideas Materials

Each DMI module consists of a Casebook, a Facilitator's Guide, and a DVD. In a DMI seminar, each participant needs a Casebook; facilitators need the Casebook, the Facilitator's Guide and the DVD. Facilitators will duplicate materials from the guide for use in each seminar session.

Casebook   The Casebook includes an introduction and eight chapters. The first seven chapters are written by actual classroom teachers detailing classroom discussions and the thinking of their students. Chapter 8 is an essay that provides an overview of the mathematical ideas of the whole seminar.

Participants prepare for each session by reading one chapter of the Casebook. Prior to the first session, participants should read the Introduction to the Casebook, as well as Chapter 1.

Facilitator's Guide For each seminar session, the Facilitator's Guide is made up of the components listed below. A description of each component follows the list.

- Session overview—an overview of the session, including how to prepare for it

- Facilitator Notes—notes that provide mathematical background for facilitators

- Maxine's Journal—a narrative account of each session from the point of view of a facilitator

- Homework responses—samples of participants' written assignments and Maxine's responses

- Detailed agenda—a thorough description of the activities of the session

- Handouts—sheets to be copied and distributed to the participants

Session Overview The session overview summarizes the main goals of the session and provides a chart suggesting the order of the activities, format, and timing for each activity. In addition, a list is provided indicating what to do to prepare for the session.

Facilitator Notes Facilitator Notes, indicated by frames, provide mathematical background for facilitators. The main Facilitator Note follows the session overview and addresses a main mathematical theme of the session. Other notes are inserted into the text where they are most relevant. The topics covered in the Facilitator Notes are

Session One
- Introducing Proof (p. 12)
- Proving the Equivalence of Two Definitions (p. 37)
- Using Algebraic Notation (p. 39)

Session Two
- Expression Generalizations (p. 52)

Session Three
- Illustrating the Commutative Property of Multiplication (p. 90)
- Thinking About $^-(^-2)$ (p. 111)

Session Four
- Using Two Models to Represent Addition of Positive and Negative Numbers (p. 119)
- Defining Sets of Numbers (p. 138)

Session Five
- Using Addition and Subtraction to Express the Relationship Among Two Quantities and Their Sum (p. 150)
- Representing a Student's Response (p. 174)
- Understanding *Identity* Elements (p. 174)
- Exploring Types of Subtraction (p. 176)

Session Six
- Examining the Distributive Property (p. 185)
- Linking Addition and Division (p. 206)

Session Seven
- Using the Term *Factor* (p. 215)
- Introducing the Notation $a \mid b$ (p. 243)

Session Eight
- Math Activity, Choice 1 (p. 270)
- Math Activity, Choice 2 (p. 272)

Maxine's Journal Maxine's Journal is a session-by-session narrative account of the RAO seminar written from the point of view of a facilitator (Maxine). It is designed to provide additional support for RAO facilitators. For each session, Maxine records small- and whole-group discussions, considers comments made by participants, and shares her thoughts, questions, concerns, and decisions. Reading Maxine's Journal as preparation for a session allows facilitators to envision the key mathematical issues that are likely to emerge, provides examples of questions that facilitators can use to drive discussion, and presents probable interactions between facilitators and participants..

Homework responses In addition, Maxine's Journal includes examples of participants' writing assignments and Maxine's responses, as well as her comments about what she was trying to accomplish with those responses.

Detailed Agendas These agendas describe each activity of a session and provide suggestions for questions that can be used to shape each discussion. When posters are to be prepared prior to a session, the agenda specifies what should be on the poster.

Some sessions include a DVD component as well, and the agenda includes a summary of the DVD cases. The DVD summaries are not full transcripts; rather, they are brief narrative descriptions of the content of the DVD segments. Facilitators should not rely solely on the narrative summaries but watch the DVD segments and take notes in preparation for leading a session.

Handouts These pages are located at the end of the agenda and are indicated by a grey strip along the edge of the page. They are to be duplicated and distributed to participants during the seminar session. The following types of handouts are included:

- Focus Questions that guide the small- and whole-group discussions of the Casebook chapters

- "Math Activities" that participants work on to deepen their mathematical understanding

- Homework pages that describe the reading and writing assignments for the sessions

- "Optional Problem Sheets" are included at the end of the Sessions 2, 4, and 6 agendas. The optional problems can be used to extend the work of the seminar and can help teachers develop additional facility with symbolic notation. Within the seminar, all participants should become comfortable with writing symbolic expressions for the ideas in the cases. The "Optional Problem Sheets" provide examples beyond those based on the cases.

DVD Segments While written cases allow users to examine student thinking at their own pace and to return as needed to ponder and analyze particular passages, the DVD segments offer users the opportunity to listen to student voices in real time. They allow participants to see students' gestures and provide rich images of classrooms organized around student thinking. These segments show a wide variety of classroom settings with diverse groups of students and teachers.

RAO Seminar Activities

In preparation for each session, participants complete a homework assignment and each 3-hour session typically consists of two or three major activities.

Homework assignments: Before each session, participants read one chapter of the Casebook and complete regular written portfolio assignments. Three times during the seminar, the portfolio assignment has participants explore their students' mathematical thinking. When participants have written such "student-thinking assignments," they share and discuss these writings with one or two other participants during the session. Other portfolio assignments ask participants to reflect on what they are learning in the seminar. Responding to participants' written assignments is an important part of the work of facilitating a DMI seminar. "Maxine's Journal" includes examples of participants' writing and a facilitator's responses.

Case discussions: In discussions of the Casebook, participants examine students' thinking, work on mathematical ideas for themselves, and consider the types of classroom settings and teaching strategies that support the development of student understanding.

Viewing DVD segments: Through the DVD, teachers can see what algebraic discussions may look like in a typical classroom setting.

Mathematics activities: Through activities designed for adult learners, seminar participants develop, share, analyze, and refine their mathematical thinking. RAO participants work on expressing generalizations in multiple ways, including words and symbolic notation, create arguments for generalizations using various representations, and discuss the connections between that work and the more formally stated Laws of Arithmetic.

Discussing the Chapter 8 essay: This discussion at the last session of the seminar creates an integrated picture of the mathematical themes under consideration, connecting the events observed in the cases and in participants' classrooms to more formal mathematics. Some facilitators prefer to assign sections of Chapter 8 during the course of the seminar instead of assigning all of them after Session 7. The agenda cover sheet for each session indicates which sections of Chapter 8 are pertinent for that session.

Preparing to Facilitate the RAO Seminar

Become Familiar with RAO as a Whole To become familiar with the flow of mathematical ideas in a seminar, we suggest that you, as the facilitator, read the introduction to the Casebook, the introduction to each Casebook chapter, and the Chapter 8 essay, "The World of Arithmetic from Different Mathematical Points of View." You should also read each session of "Maxine's Journal" to get an image of the seminar experience from the point of view of a facilitator. In addition, you may find it useful to examine some of the cases and look through the detailed agendas and handouts.

Identify Connections Between Goals for the Seminar and Goals for Each Session Once you are familiar with the goals and components of the RAO curriculum as a whole, the next step is to prepare for individual sessions. For each session, read the cases, the related entry of "Maxine's Journal," and the agenda. Work through the activities in the session yourself; for example, work through the math activity, view the DVD segment, think about the "Focus Questions," and familiarize yourself with any additional handouts. As you do this work, think through the issues raised by that set of activities. What are the goals of the session as a whole? What ideas about mathematics learning and teaching should emerge as teachers participate in the investigations and discussions? How are these ideas illustrated in the cases? How might they arise in the other activities? What questions might you pose to call attention to these ideas?

Prepare the Logistics In addition to planning for issues likely to arise during discussions, you must think through the order of the activities and review the suggested timetable. Organize the readings, handouts,

DVD, DVD equipment, and manipulatives so that you are able to focus on the seminar participants during the session. Suggestions for time allotments and the order of activities are given in the session agendas, along with lists of the materials you will need for each session.

Comments from Other Facilitators

Those who field tested the RAO seminar had recommendations for other facilitators. Several commented about planning:

I realize now more than ever how important it is to be really prepared and to have thought through the issues, mathematical and otherwise, that might arise. Having a sense of the important points that you want people to be exploring and the direction in which you want them to be headed is crucial. However, it is important to realize that sometimes "the way there" might turn out to be different from the route you anticipate.

In conducting the RAO seminar, it is important to keep in mind both the goals for the entire seminar and how the goals for each individual session fit together to make a coherent whole. One facilitator suggested the following:

Make an overall outline of the entire course. Each time you plan a session, consider how that session fits into this outline so you can sense how the ideas are building.

Several field-test facilitators offered suggestions about how to become familiar with the mathematics of the RAO seminar and recognized the power of building models during the seminar. One of those is shown below:

Work through the mathematics before the session, taking the time to build several models and write a variety of story contexts.

This will be beneficial as you facilitate discussions about the relationship between the models, the context, and the symbols. It is very difficult to anticipate what the participants might come up with or to help them make connections if you haven't done this yourself.

Some facilitators were concerned about whether their background in algebra was strong enough to address issues that would arise in their seminar. They found that it was helpful to have identified a person familiar with formal mathematics whom they could go to with questions between seminar sessions. In the words of one facilitator:

Identify a person who has experience with algebraic thinking that you might use as a resource when issues arise in your seminar that you haven't yet had time to think about yourself. When issues arise that participants want to know more about, but you haven't thought about, give yourself permission to say that all of you can think further about those issues for the next session. You can decide if the issues are essential to discuss with the whole group at a later session or whether they can be addressed with a few people during a break.

Some facilitators wrote about how to address participants who had preconceptions about the role of models in a mathematics class. One comment follows:

During the sessions, insist that participants develop models to illustrate the ideas of the sessions. Models assist in mathematical communication, but they aren't just about "sharing your ideas." Rather, the model brings concrete expression to the idea. Participants might come to the seminar thinking the models are for the less adept students or that they are ways to demonstrate that you understand. Coming to see that creating models is an important part of mathematical reasoning is a goal of the seminar.

Another wrote about how to help elementary teachers see the connections between the algebra they are learning in the seminar and the computational goals they have for their students:

It is important for the participants to see the connection between algebraic reasoning and computational fluency. Since many elementary teachers are just beginning to see the value of

exploring algebraic ideas with their students, this link gives them a good starting point for the rationale and importance of providing their students these opportunities.

Another facilitator offered advice about how to help participants make connections among the various activities in the seminar:

Have posters displayed around the room to keep the goals of each session in the forefront. Continue to make connections to those goals during each experience of the session.

First Homework

This assignment is to be completed before the first session of the seminar. In that first session, you will have the opportunity to share your thinking about algebra with other participants. You will also discuss the Casebook readings in that session.

Writing assignment: What does algebra mean to you?

Write a reaction to each of the following statements in one or two paragraphs:

■ When you hear the word *algebra*, what kinds of mathematical ideas come to your mind?

■ What, if anything, does algebra have to do with the content you teach?

■ What might it mean to engage with children on algebraic ideas?

Reading assignment: Starting the Casebook

In preparation for the first session, read the introduction to the Casebook, *Reasoning Algebraically About Operations*. Then read Chapter 1, "Discovering Rules for Odds and Evens," including both the introductory text and Cases 1–5.

Maxine's Journal

September 15

Preseminar reflections

I'm about to start facilitating a new DMI seminar based on ideas of early algebra: Reasoning Algebraically About Operations. We'll be looking at students' ability to make generalizations about how the operations work, generalizations that, in later years, could be written in algebraic notation. For now, we are not interested in having students use such notation. Instead, we want to know what happens when students are encouraged to articulate their observations, challenged to consider whether what they notice will *always* work, and asked to prove this. As students start to notice such generalizations about numbers and operations, they take their first steps from thinking about operations exclusively in terms of actions to also thinking in terms of a system of relationships.

I expect that thinking about generalizations and proof will be new for many of the seminar participants. In this setting, I am not so interested in formal mathematical proof in which one begins with a set of axioms from which all other theorems are derived. Rather, I want the whole group to work with a variety of representations of the operations to help them understand if the relationships they see will always hold. This kind of investigation is what we could expect in elementary- and middle-school classrooms.

I also have mathematical goals for teachers that extend beyond interpreting the thinking of K–8 students. I want teachers to explore integers and be able to make sense of rules for ordering, adding, and subtracting integers. By the end of the seminar, I want them to be familiar with the nine Laws of Arithmetic and understand their importance. I also want teachers to become more fluent with algebraic notation. However, I feel I need to be very careful with this last goal. I know many teachers have had unpleasant experiences with algebra in the past, and the sight of algebraic symbols induces panic. I do not want to push the participants toward the use of algebraic symbols if doing so causes them to lose confidence in their own ability to think through mathematical ideas.

I have set as a prerequisite for this course that teachers have already participated in the DMI seminar, Making Meaning for Operations, or a course that addresses similar content. I feel that is crucial for what lies ahead of us in this seminar. Before we begin the work of Session 1, teachers must already have thought about what the operations mean and how they can be represented with cubes, pictures, and story problems. They should already be familiar with representing addition and subtraction on number lines, and with representing multiplication and division with arrays or rectangles. In addition, it is important that participants recognize how numbers can be decomposed to facilitate computation and how the parts are put together again for the different operations.

Because everyone in this seminar has already participated in DMI seminars, I can assume (unlike other seminars I've taught) that they each recognize they have their own mathematical thoughts, as do their students. I also assume that they are practiced in analyzing students' ideas and looking for the logic in what a student says and does, even though there may be flaws in the student's thinking. In addition, I assume that they have experience with reading and discussing cases, working on mathematics activities in small groups, and participating in whole-group discussions.

Even so, I expect we will need to work to come together as a group. The seminar will be meeting every one or two weeks from 4:15 P.M. to 7:15 P.M. I have the list of participants who have signed up and a handful of them have taken DMI seminars with me before, though most have studied with other facilitators. Most of the participants are teachers, one is a principal, and two are math coaches. Everyone has been sent packets of materials in order to prepare for this course. For our first meeting, they have been asked to read the introduction and the first chapter of the Casebook, write about what they think algebra is, and respond to some questions that will be used as a preseminar assessment.

REASONING ALGEBRAICALLY ABOUT OPERATIONS

Discovering Rules for Odds and Evens

Mathematical themes:

■ Are two different definitions of even numbers equivalent?

■ What comprises an argument that a statement is always true when you cannot check every number?

■ What are generalizations about adding and multiplying odd and even numbers and how can they be proved?

Session Agenda

Sharing homework (ideas about algebra)	Groups of three	10 minutes
Introductions	Whole group	10 minutes
Definitions of even	Small groups	15 minutes
Definitions of even discussion	Whole group	10 minutes
DVD: Students talk about even numbers	Whole group	15 minutes
Math discussion: What is a mathematical argument? Adding odd numbers	Partners	5 minutes
	Whole group	15 minutes
DVD: Adding odd and even numbers	Whole group	10 minutes
Break		**15 minutes**
Math activity: Operating with odd and even numbers	Small groups	20 minutes
	Whole group	20 minutes
Case discussion: Students think about odd and even numbers	Small groups	15 minutes
	Whole group	15 minutes
Homework and exit cards	Whole group	5 minutes

Background Preparation

Read

- the Casebook, "Introduction" and Chapter 1

- "Maxine's Journal" for Session 1

- the agenda for Session 1

- the Casebook, Chapter 8: Sections 1, 2, and 3

Work through

- the Math Activity, "Operating with odd and even numbers" (p. 46)

- the Focus Questions for Session 1 (p. 47)

Preview

- the two DVD segments for "Students talk about even numbers" and choose which segment to show

Materials

Duplicate

- "Math Activity: Operating with odd and even numbers" (p. 46)

- "Focus Questions: Chapter 1" (p. 47)

- "The Portfolio Process" (p. 48)

- "If You Have to Miss a Class" (p. 49)

- "Second Homework" (p. 50)

Obtain

- DVD player

- cubes

- index cards

Prepare two posters

- Hue and Julio scenario (p. 45)

- Criteria for Representation-Based Proof (p. 45)

Introducing Proof

In mathematics, a *theorem* must start with a mathematical assertion, which has explicit hypotheses (or "givens") and an explicit conclusion. The proof of the theorem must show how the conclusion follows logically from the hypotheses. For instance, in Case 4, Amanda asserts that if two numbers are even, then their sum is an even number. In later years, Amanda's theorem might be stated as: If two numbers, m and n, are both even, then their sum, $m + n$, is an even number. The Facilitator Note, "Using Algebraic Notation," illustrates a proof of this claim. This proof consists of a series of steps that begin with the hypotheses, that m and n are even numbers, and forms a chain of logical deductions ending with the conclusion, that $m + n$ is an even number. Each deduction is justified by an accepted definition, fact, or principle.

One does not expect the rigor or sophistication of a formal proof, or the use of algebraic symbolism, from young students. Even for a mathematician, precise methods of validation are often developed *after* new mathematical ideas have been explored and are more solidly understood. When mathematical ideas are evolving and there is a need to communicate the sense of *why* a claim is true, then other methods of proving are appropriate. Such methods can include the use of visual displays, concrete materials, or words. The test of the effectiveness of such a justification is: Does it rely on logical thinking about the mathematical relationships rather than on the fact that one or a few specific examples work?

Throughout RAO, participants work to prove general claims (or theorems) about an infinite class of numbers using visual displays, concrete materials, or story problems. The criteria used to determine whether a representation constitutes a proof of the general claim are:

1. The meaning of the operation(s) involved is represented in diagrams, manipulatives, or story contexts.

2. The representation can accommodate a class of instances (for example, all whole numbers).

3. The conclusion follows from the structure of the representation.

For example, consider Madelyn's proof (see "Maxine's Journal," p. 20) showing that the sum of two odd numbers is even. She holds up two collections of cubes, each consisting of pairs plus a single cube. Then, she shows how these two collections represent *any* two odd numbers by covering most of the pairs in each collection with a sheet of paper.

Discovering Rules for Odds and Evens

She explains that when the two collections are joined, the two single cubes come together to make a pair. Now the total is represented by a larger collection of pairs of cubes, so the total is an even number.

In Madelyn's representation, the operation of addition is represented as the joining of two sets. By explaining that each set can be seen as any number of pairs plus a single cube, any positive odd number is represented. (If she were to move her paper to hide the one pair showing, her representation would include 1.) And the representation shows how the conclusion—that the total is even—follows from the premise that two odd numbers are added.

Note that Madelyn's proof applies only to positive numbers, the realm in which students are working, whereas the algebraic proof shown in the Facilitator Note, "Using Algebraic Notation," applies to integers, including negative numbers and zero.

Maxine's Journal

September 17

The session last night certainly was jam-packed. The agenda for the DMI sessions is always full, but last night, in addition to handling all of the introductory logistics at the beginning of any seminar, there were several big ideas to move into. We started thinking about making generalizations, proving generalizations with visual representations, and exploring how students move into these ideas. All of that was done in the context of odd and even numbers—teachers working on the mathematics for themselves and also examining students' thinking through print cases and a DVD segment. I think it went well. At the end, it was clear there was a lot to think about. At this point, ideas about stating and proving generalizations aren't solidly held, but that is appropriate at the beginning of an eight-session course. Everyone seemed to be engaged, and though some seemed puzzled, they all seemed to leave in a good mood.

Before the session, I posted the agenda so that people would know what to expect. I also wanted to communicate that we would begin each meeting promptly at 4:15 p.m. and would end at 7:15 p.m.

Sharing ideas about algebra

As the session began, I put the participants into groups of three and asked them to introduce themselves to one another and to share their experiences with algebra. A few people came in late, and I settled them into groups as they arrived. Then I had a few minutes to listen in on conversations. There were several people who said they enjoyed algebra when they were in school; they thought of it as a puzzle and liked being able to find the correct answers. There were also participants who said they hated algebra and maneuvered the rest of their education to avoid taking math classes whenever possible—until they became educators involved in professional development, that is. Some people said they were able to get good grades in algebra, but they never felt like they truly understood it.

Vera, the principal, said that she didn't have good experiences in algebra and had never actually had good experiences in mathematics until she started taking DMI seminars. This is her fourth seminar and she felt it was important to learn about algebraic thinking because, as a K–12 principal, she wants her teachers to recognize the importance of helping students think algebraically.

After 15 minutes, I called the group together and asked that each participant think of one word or phrase that relates to his or her experience with algebra. Then, I had them introduce themselves and share their comment with the group. I did not want to spend a lot of time on this whole-group discussion, but I did want to give all participants an opportunity to share their thoughts. This is the list that resulted:

- Patterns
- Formulas
- Variables; x and y
- Scary
- Using a known to solve unknowns
- Equations
- Bad thoughts
- Symbols instead of unknowns
- Fun
- If you do it to one side . . .
- Balance a scale
- Functions
- Missing addend problems
- Generalizations
- Thinking
- Yuck
- Satisfying
- High school
- Puzzle like
- No understanding of why

I thanked the group for sharing these impressions. I mentioned that we would discuss logistics later in the session, but, at this point, I wanted to continue with our discussion about algebra.

Definitions of even

I began by saying a few words to connect what we had just been talking about to what we were about to undertake for the next eight sessions.

"I found it very interesting to hear your words and phrases, and I am looking forward to reading what you have written. I want to make it clear that there is not any fixed definition of algebra. A few years ago, I read the proceedings from a conference of mathematicians and mathematics educators who came together to discuss what algebra is—and they never came to an agreement. The one thing they did agree on is that there is more to algebra than x's and y's; that is, there is more to algebra than just the notation. Those x's and y's have to do with ideas, even though many people go through an algebra class without any sense of what those ideas are. In this seminar, we'll be looking at what the ideas are that can be expressed using x's and y's, but we'll be focusing more on the ideas than on the notation.

"The title of the seminar is Reasoning Algebraically About Operations. Throughout the seminar, we'll be looking at what it means to think about numbers and operations as a system. We'll be thinking in terms of generalizations about how the operations work and how the operations are related to each other.

"Today we will be thinking about odd and even numbers and generalizations that can be made about what happens when we operate on them. These ideas *can* be expressed in terms of x's and y's, but as you read in the cases, students have ways of thinking about odd and even numbers without reference to x's and y's. We will get to the case discussion in a little bit. To start, however, we are going to think about how to define even and odd numbers."

I put up a poster that presented the following scenario:

> In a second-grade classroom, the teacher commented there was an even number of students in the class that day. Hue said, "I knew it is even because when we lined up for lunch everyone had a partner." Julio said, "I knew it was even because when we split into two groups, the two groups were equal and no one was left over."

Small groups

I asked people to work in small groups to do two things. First, I wanted them to write out the definitions of even number implied by Hue's and Julio's statements. Then, I asked them to consider whether these definitions of even numbers described the same set of numbers and to explain how they know.

In small groups, the participants thought through the difference between Hue's and Julio's observations. June mentioned she thought Hue was thinking about skip-counting, but Julio was thinking about division. Yanique pointed out that Hue's and Julio's observations could both be thought of as either multiplication or division. Then she asked, "Don't they each illustrate a different way of thinking about division?" When I asked her to explain, she said that Hue thought of the class divided into groups of two and Julio, two groups.

Whole group

After 15 minutes, I pulled the participants together into a whole group to share their thinking about these questions. We wrote two definitions on the board:

- An even number can be made into pairs (groups of two) with no leftovers.
- An even number can be split into two equal-sized whole-number groups with no leftovers.

Initially, Jorge suggested a slightly different version of the second definition, "An even number can be split into two equal-sized groups." However, June pointed out that would make 5 an even number because, after all, it can be split into two groups of $2\frac{1}{2}$. So, we decided to change Jorge's definition to address that issue.

When I taught this seminar the first time, I was surprised to discover that many people—students *and* teachers—didn't realize that the definitions of

Discovering Rules for Odds and Evens

even and odd were based on the number 2, and so I underlined *pairs* and *two* in the definitions. I mentioned that, even though we talk about a number dividing another number *evenly*—for example, 3 divides 9—that does *not* mean 9 is an *even number*. I suggested they keep an eye out for this confusion.

Then I asked participants how they knew these two definitions always describe the same set of numbers.

Grace held up a set of cubes arranged as an array two cubes high. "If you look at an array, you can see why it has to be that way. If a number can be split into two equal-sized groups of whole numbers, you have got these two equal rows. However, then you can look at the array the other way and see all the columns of two."

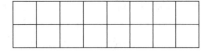

Kaneesha said, "This is what I once heard a student say: 'You have people on two soccer teams. On each team, everyone has a position—a goalie, a right forward, a left forward. So each person has a partner in the same position on the other team. If you have two equal teams, everyone has a partner.'"

Josephine held up a bunch of cubes arranged in an irregular configuration and waved them around so nobody could get a fix on them. "So I want to know, is this an even or an odd number?"

I was not sure if Josephine was really asking a question, but there was an important point to be made from what she was doing. "Josephine is pointing out that, if the cubes are in a different configuration, it's hard to tell if they represent an even number or not. The thing is, if it's even, it's even, no matter what the configuration. It's just hard to tell the way she has them arranged right now."

We spent just a couple of minutes on the definitions of odd numbers. We came up with several variations of a definition:

- An odd number, when put into pairs, has one left over.
- An odd number, when divided into two equal whole-number groups, has one left over.
- An odd number is an even number plus one.
- An odd number is an even number minus one.

DVD: First and second graders talk about even numbers

Of the two DVD segments on definitions of odd and even numbers, I chose to show the one of a first- and second-grade combination class. I said, "In this class, students define even numbers as those that can be divided into two equal whole-number groups. However, the teacher, Karen, noticed that some of her students started answering questions in terms of whether the number could be divided into pairs. This is what she wanted to investigate with the class."

I showed the first portion of the DVD, up to where Jesse points out that the two towers of cubes can be seen as two and two and two and two.... Then, I asked the whole group what they saw.

Grace pointed out that students looked at the cubes the same way she did. If the number of cubes is even, the cubes can be arranged into two towers, and then you can look across the towers to see them in pairs.

I asked if people agreed, and they nodded. However, then Madelyn brought up a different issue. "When I hear those students and see the way those cubes are arranged, I wonder if their idea of even is that the cubes are 'level.' Like, I do not know if they get Josephine's point—that evenness has to do with the number, not the way the cubes are arranged."

Denise, a math coach, agreed with Madelyn and then raised another question. "I wonder if they get the importance of the number 2. Like what you said before, I see in some of the classes I've been visiting, some of the students were thinking that an even number is any number that can be divided evenly. Like, they would say 15 is even because you can make 3 even groups of 5."

The issue that Madelyn and Denise were raising is an important one about the common misinterpretations of the definition of an even number based on how the cubes are arranged. I paused here to have some members of the whole group paraphrase the ideas before I asked, "What questions would you pose to students to get at these issues?"

Denise said that she would show the students three towers of five and ask if they represent an even number. Madelyn said she would show the students an even number of cubes divided into two uneven groups and ask if those represent even numbers. If students answered these questions incorrectly, then the teachers would know that there is a concept that they need to address.

I suggested that we look at the last minute of the DVD segment to see if the teacher, Karen, addresses their concerns. In this segment, Karen shows the class two towers, one of which is one cube higher than the other. The class agrees this is an odd number. Then, Karen takes one cube off the shorter tower. "What about this?" she asks. At first, some of students call out, "Odd," but then, after a pause, a student says, "No, even," and soon they all declare it is even. To illustrate, one child starts breaking the towers into pairs.

After viewing this segment, the participants agreed that Karen's question and the class response showed that at least some of students knew that *even* doesn't mean "level." One participant commented, "We don't really know how students who first said it was odd were thinking." Denise also said that she was not really convinced that if the teacher held up three towers of five, some of the students would not say the amount was even.

I told the group that, for homework, they would be in the position to ask such questions of their students. However, we would discuss the homework later. For now, I wanted to turn to their own exploration of odd and even numbers.

Discovering Rules for Odds and Evens

Math discussion: What is a mathematical argument? Adding odd numbers

I continued, "We are going to start off thinking about what happens when you add two odd numbers. When you add 3 + 5, two odd numbers, you get 8, an even number. Do you think that always happens?"

When James said, "Of course," I asked, "How do you know?"

"Well, just try it: 5 + 7 is even; 3 + 11 is even; 23 + 17 is even. Add any two odd numbers and you will see, the answer comes out even." | 195

Of course, James is right. If you add any two odd numbers, the result will be even. However, I wanted the participants to be able to think about a mathematical argument for this concept. I wanted them to think about how to explain *why* it always works out that way and how they can be absolutely | 200 sure that there will not be an exception.

To illustrate the importance of finding another form of argument, I proposed we look at a different statement for a moment. "Consider this: The sum of two prime numbers is even. All of James's examples fit this statement: 5 + 7, 3 + 11, and 23 + 17. They are all examples of adding two prime numbers and getting | 205 an even sum."

Jorge said, "Well, maybe it is true."

Denise said, "It is not true. Look at 2 + 5; 2 and 5 are both prime numbers, but 7 is odd."

I said, "That's right. The statement is not true. However, if you didn't think | 210 to try 2, you could have gone on testing out prime numbers and only finding examples that fit the statement. So we are going to be looking for other arguments to prove our statements, ways that show us *why* the statement has to be true. Let's go back to the idea of adding two odd numbers. Does anyone have a way of thinking about it more generally?" | 215

Leeann said, "Two odds make an even because two negatives make a positive."

At first, I was somewhat dismayed to hear this. What was the connection that Leeann was making? We were not talking about negatives at all. Besides, when you *multiply* two negative numbers, you get a positive, but we were | 220 talking about *adding* odd numbers.

Then I realized that Leeann's comment represents an important issue of this seminar. That is, sometimes people present images or metaphors that have some vague connection to the idea under discussion and think it's sufficient as an explanation or a proof. I suppose Leeann was thinking about two | 225 of one kind of number making another kind of number. It's my job to try to clarify what is entailed in making a mathematical argument.

I asked Leeann to explain what she meant, and she said, "Like, if I say I am not not hungry, that really means I am hungry."

So, Leeann was not thinking about numbers at all! I said that was an interesting metaphor, but it does not really explain how we know the sum of two odd numbers is even. "What we are looking for is an explanation that includes what an odd number is and also shows us the action of addition."

Charlotte said, "Yeah, I don't think of an odd number as a negative. After all, there is that 'plus 1' in there."

Then, Madelyn got us on the right track. She said, "I was thinking about the way some of the students in one of the cases were looking at odd numbers. They showed 5 + 5 like this." Madelyn held up two groups of five cubes and showed what happens when they are brought together.

"When you bring the two 5s together, the extra cube that is hanging out in one of the 5s matches up with the extra cube that is hanging out in the other. So then all the cubes are paired up, and you have an even number."

OK, now we were moving in the right direction. Madelyn's model had a representation of odd numbers and a representation of even numbers, and it embodied the action of addition. At this point, her explanation was about specific numbers—5 and 10—so there was more work to do. First, I wanted to make sure everyone understood what she showed us.

I thanked Madelyn for her demonstration and asked someone else to paraphrase and enact what she did. After Leeann complied, I asked if there were questions. When there were not, I said, "Now, Madelyn and Leeann have shown us what happens when you add 5 + 5. How does that help us think about adding *any* pair of odd numbers?"

Madelyn was ready to jump in, but I gestured that she wait to give everyone time to think about the question. "We want to show that, no matter what two odd numbers you start with, if you add them together, the sum will be even. How do we know that always works?"

After a pause, I let Madelyn continue with her explanation. "I happened to show it with two fives, but it could have been any odd number. I can cover up the cubes so you don't know how many there are." She asked Phuong, who was sitting next to her, to hold up the two groups of five cubes. Then, Madelyn partially occluded the cubes with sheets of paper.

"I can tell you that there are some numbers of cubes behind the papers. You do not know how many there are, but I can tell you that they are all paired. Because the cubes behind the papers are paired, then the ones hanging alone that you see make both sets odd numbers. When you join the amounts, it still doesn't matter how many cubes are paired. The two cubes

Discovering Rules for Odds and Evens

that are hanging out get paired, and so all the cubes are paired, and you have an even number."

At that point, I asked everyone to talk through Madelyn's argument in small groups. What she demonstrated was key to understanding the kind of mathematical arguments we would be working with during this seminar. Madelyn used a representation that characterizes odd and even numbers; the action on the representation shows the operation of addition as two quantities joined; and the representation can illustrate *all* odd numbers. As the representation embodies all of these things, it shows us that no matter what two odd numbers you add together, the result *always* has to be an even number.

After everyone had a chance to work on this idea in a small group, I asked Yanique to talk us through the argument again. When she finished, I pointed out the difference between an argument like this one and the one James offered earlier in the discussion. "In mathematics, there is a kind of skepticism. If you rely on testing things out a bunch of times, someone can always say, 'Well, maybe we just have not come across the ones that don't work. There might be some numbers out there that don't work.' But, when we rely on a representation that shows us how things are related, we know that it always has to work that way."

Then I posted the following chart:

Criteria for Representation-Based Proof:

1. The meaning of the operation(s) involved is represented in diagrams, manipulatives, or story contexts.
2. Representations accommodate a class of instances (for example, all whole numbers).
3. The conclusion follows from the structure of the representation.

I explained that when we considered proofs in the elementary classroom, this is what we would be looking for.

Charlotte asked why the chart said "Representation-Based Proof" rather than just "proof." I explained that mathematicians generally use other criteria when proving general claims about the number system. Specifically, they rely on the Laws of Arithmetic as their starting point. However, because elementary- and middle-school students typically do not have an understanding of the Laws of Arithmetic, we need to begin somewhere else. So we are using representations of the operations—what the operations mean—in order to show how we know something *has* to be true. I continued, "Using representations, such as what Madelyn showed us, allows us to make a claim about *all* odd numbers. Testing a bunch of specific examples doesn't allow us to make such a claim."

Leeann asked how Madelyn's proof showed the meaning of the operation. I pointed out that the operation is addition, and joining the two sets that represent odd numbers shows it.

Grace said, "I had a different way of showing that the sum of two odds is even, and I think it works for any pair of odd numbers. If you have any odd number of cubes, you take off one cube and it is even. So if you have two odd numbers, you take off the one cube from each and put them together. Now you have an even number plus an even number plus 2, and so the answer is even." As Grace talked, she showed us what she was doing with cubes.

"Yes, that's a good argument," I said. "The thing is, you rely on something we have not yet proved. You said that the sum of even numbers is even. Now, that might seem obvious, but we have not yet shown why that is the case. Let's go through the argument that the sum of even numbers is even."

Grace was ready to continue, but I wanted to get someone else involved in the discussion, so I waited for more hands to go up and called on Antonia. She explained, "If you have two even numbers, each can be shown as some number of pairs. When you add, you join the two amounts, and they make up a bigger number of pairs, but they're still pairs."

Grace was nodding as Antonia spoke. I pointed out that now, with that explanation, Grace's argument is another way to show that any two odd numbers total to an even number.

DVD: Adding odd and even numbers

At this point, I started the DVD again to show a group of fifth graders working on addition of odd and even numbers. When asked about the sum of three odd numbers—9, 11, and 7—some students used representations to explain how they knew the sum would be odd without actually doing the calculation. The teacher, Janelle, has not asked her class whether the sum of three odd numbers would *always* be odd, but students' demonstrations could easily be extended.

Time was slipping by. Once the participants watched the DVD, I called a break. We would refer back to the DVD as participants worked on their arguments in the next activity.

Math activity: Operating with odd and even numbers

As I handed out the "Math activity" sheet, I told the group that we would now be thinking about what happens when you *multiply* odd and even numbers. "Again, I want you to figure out what happens when you multiply two odd numbers, or two even numbers, or an odd number and an even number. Then I want you to come up with an argument that shows how you know that will *always* happen."

Small groups

A few minutes into the work, Risa called me over. She explained that her group—Phuong, Leeann, and Risa—was still thinking about addition and had some questions for me. Looking at the materials in front of them, I saw that Phuong used graph paper for her arguments and Leeann used cubes. Both representations showed that when you add two even numbers, you join a bunch of pairs with a bunch of pairs, and you end up with a larger bunch of pairs. They were clear that, although they used different materials, their arguments were essentially the same.

Risa, however, was using algebraic symbols. "See," she said, "An even number is written as $2n$. So you've got $2n + 2n$, which is $4n$, and that is even, too."

I said, "Yes, in algebraic notation, the definition of an even number is a number that can be written as $2n$, where n is an integer. Because we are working with whole numbers right now, let's just say that n is a whole number. However, when you use $2n$ for both of your even numbers, you're saying that they are the same number, like adding $6 + 6$. If you want to show that the two numbers might be different, you need to use a different letter. So you might say $2n + 2m$, where n and m are both whole numbers. Then, when you add them together, you get $2(n + m)$. Because $n + m$ is another whole number, that means $2(n + m)$ is even."

Risa nodded, indicating that this was clear for her. Now, I wanted her to be able to see how her algebraic notation relates to Leeann's cubes and Phuong's graph paper. "When you look at your notation, you can read $2n$ as saying you have n pairs; $2m$ means you have m pairs. Now look at your partners' representations."

Phuong stepped in, "My representation shows n pairs here and m pairs there. When you put them together, you have $n + m$ pairs. You can see it in the cubes and on the graph paper, and you can see it in Risa's equation."

Leeann laughed and said that the symbols make her jittery. I told her she had a good way of thinking about these arguments, and I wanted her to keep working on her ideas. At least for now, I'll work on the algebraic notation only with those participants who bring it up. I want to encourage participants who become nervous when they see algebraic symbols to concentrate on developing their arguments using cubes or other representations that they feel more comfortable with.

When I got to Antonia, June, and Josephine, they said they had worked on the rules for multiplying odd and even numbers. However, Josephine had another question. "When I learned about even numbers, I was told they are numbers that end in 2, 4, 6, 8, and 0. That works. All you have to do is look at the last number."

I started to question Josephine, and in response to each question, she said, "It works. It always works." I realized she did not understand my questions, and I needed to use another approach. I said that she was right; I agreed that this always works. Then I asked her, "How does that fit with what we are learning about even numbers? How do we know that if the ones digit of a number is even, then it will fit with our definitions: 1) that the entire number can be divided into two equal whole-number groups and 2) that the entire number can be divided into pairs?"

Once I acknowledged that what Josephine was saying about even numbers is true, she seemed to be able to hear the next question. However, it was really the other two participants in her group who took on the challenge. June said that any number is a multiple of 10 plus the number in the ones digit. "Because we know that an even plus an even is even, that means we need to prove that any multiple of 10 is even." Antonia said, "Because 10 is even, if you add up any number of 10s, you'll have a bunch of pairs. That is what you need to prove it."

I said, "OK, any multiple of 10 is even. Now explain to me how you know that the ones digit determines whether a number is even or odd."

June finished the argument. "We said that any number is a multiple of 10 plus the ones digit. If the ones digit is even, then you have an even plus an even, which is even. If the ones digit is odd, then you have an even plus an odd, which is odd. So there it is."

They looked at me for confirmation. "Sounds good to me," I said.

Other groups I visited were working on the questions about multiplication. Jorge, Grace, and Vera had made up a multiplication table, which they were color-coding. It seemed like a good way to look for patterns in what was occurring. Later, I would try to get back to them to see how they were thinking about their proof.

	1	2	3	4	5	6	7	8	9	10
1	1	2	3	4	5	6	7	8	9	10
2	2	4	6	8	10	12	14	16	18	20
3	3	6	9	12	15	18	21	24	27	30
4	4	8	12	16	20	24	28	32	36	40
5	5	10	15	20	25	30	35	40	45	50
6	6	12	18	24	30	36	42	48	54	60
7	7	14	21	28	35	42	49	56	63	70
8	8	16	24	32	40	48	56	64	72	80
9	9	18	27	36	45	54	63	72	81	90
10	10	20	30	40	50	60	70	80	90	100

When I visited Lorraine, Claudette, and Mishal, they were looking at an arrangement of four groups of three cubes. Claudette said, "The 4 makes it even. If it were 5 groups of 3 it would be odd." Mishal added, "It has to be an odd number times an odd number to be an odd number." 410

I asked, "How do you know that?"

Mishal explained, "If you have an even number of groups, you can pair them up." She showed me how her four groups of three could be seen as two groups of three and another two groups of three. "You know that each pair of groups is an even number. It's a number divided into two equal groups. When 415 you add up all those even numbers, the total is even."

Claudette interjected, "And if you have any number of even groups, you know you have an even number. Again, all the cubes in those groups can be paired. When you put them all together, you have a whole slew of pairs."

Mishal then finished the argument. "If you have an odd number of groups, 420 and the groups are odd, you pair off as many groups as you can, and that makes an even number. Then there is one odd group left. You take as many pairs as you can, but there is one left over. So in the whole amount, you have all those pairs and one left over, so it has to be odd."

I looked over at Denise, Kaneesha, and Yanique. They had made towers of 425 five and were showing what happens each time they add a group. "One group is odd, two is even, three is odd, and four is even." They looked at me, and Denise said, "When we did it with an even number, the total was just even, even, even. However, when we keep adding on groups of odd numbers, the total becomes odd, even, odd, even." 430

Whole group

I never did get a chance to check back with Jorge's group, but I decided to start the whole-group discussion by addressing that group because I wanted the class to see what Jorge, Grace, and Vera had done with charts. It turned out that they had spent all their time making pages of multiplication charts up to 50 × 10. The three members of the group proudly showed their work, explain- 435 ing that it was exciting to see the pattern emerge. Grace declared, "Anything times an even number is even. An odd times an odd is odd."

I nodded and acknowledged the careful work they had done. Then I asked the rest of the participants whether this work convinced them that an odd number times an odd number is always odd. For a moment, everything was 440 quiet. Then Risa said, "The charts give me a sense of the structure of the number system, and it makes sense that the pattern would continue."

M'Leah ventured, "It seems like the pattern will continue. It's sort of like giving a bunch of examples, but it doesn't really explain why the pattern works." 445

I asked whether anyone would like to offer an explanation for why the pattern works that way.

James said, "Lourdes has a cool way of looking at it."

I hadn't made it to Lourdes's group and didn't know what she would offer, but indeed, her way of thinking about it was nice.

Lourdes said, "I started thinking about 5 + 5 + 5. Really what I need to think about is the 1 in each of them that makes the number odd. So at first I wrote it like this:

$$5 + 5 + 5$$
$$\bullet \quad \bullet \quad \bullet$$

"I put a dot under each 5 to show the extra 1. Then, James pointed out that instead of the 5 it could be any odd number. On top of the dots, you can imagine there are any number of pairs."

June said, "Wait a minute. Are we talking about multiplication? Lourdes added."

James responded, "She is using repeated addition. That is multiplication, isn't it?"

I acknowledged that whole-number multiplication could be represented as repeated addition.

June said, "Oh, yeah."

When I asked if there were any other questions about what Lourdes had done, Risa said, "I don't get what the dot is if the number is 5."

Phuong commented, "It's just like what the fifth graders on the DVD did."

Lourdes explained that the dot stands for the 1 that is left after you put things in pairs. "It is sort of like what students on the DVD did, and it is also like what Madelyn did before, holding the paper up so we couldn't see all the pairs. Only I didn't put a paper up. You can imagine the pairs are there; I just didn't draw them. The number does not have to be 5. The number could be 3 or 7 or 25 or 31. That dot is still going to be there after the rest of the number is in pairs."

Once it seemed that everyone understood Lourdes's representation, I said, "It looks to me as though Lourdes has shown us that 3 times any odd number is odd. Is she finished?"

Antonia asked, "Doesn't she still need to show us that any odd number times any odd number is odd?"

As Lourdes had already sat down, I asked if anyone else would like to finish the argument. Denise rose, erased Lourdes's 5s, and replaced them with x's.

$$x + x + x$$
$$\bullet \quad \bullet \quad \bullet$$

"I'll say that x stands for any odd number. Like Lourdes said, the dot underneath shows what is left after you have taken out all the pairs. So if you have 3 times x, you have 3 dots. If you take out another pair of dots, you are left with 1 dot, which means you have an odd number. If you add another x, you also add another dot.

$$x + x + x + x$$
$$\bullet \quad \bullet \quad \bullet \quad \bullet$$

"Now you have 4 times x, an even number times an odd. You can pair up all the dots, which means you have an even number. You can keep going like that. If you have an even number times x, you end up with an even number of dots, so the answer is even. If you have an odd number times x, you end up with an odd number of dots, so the answer is odd."

I asked if there were any questions for Lourdes, James, or Denise. Mishal said, "I think Lourdes's way is pretty much like what we were thinking, except we did it with cubes instead of dots. Also, we first showed the multiplication and then we thought about the pairs and whether there would be an extra cube. Lourdes separated out the pairs and the 1 from the beginning."

Once the group was satisfied with Lourdes's explanation, I emphasized what made it a justification for why the product of two odd numbers is odd. "First, as June highlighted for us, the representation shows multiplication as repeated addition." I pointed to the chart with the Criteria for Representation-Based Proof. "That satisfies the first criterion. Lourdes demonstrated a way to show any odd number. She showed us the dot and said that each dot represents the 1 that is left after everything else is in pairs. Denise has shown us how to look at the representation and extend it to any number of odd numbers. So the second criterion is satisfied. Also, the representation shows that, given the premise that you multiply two odd numbers, the conclusion follows—the product is odd. That is, the conclusion follows from the structure of the representation, not from a particular instance, and so it satisfies the third criterion."

Case discussion: Students think about odd and even numbers

There were several pretty complex ideas in the cases that I wanted participants to address. First, I wanted them to recognize the importance of students' generalizations about the number system and start noticing when, in fact, students are generalizing.

Second, I wanted participants to understand the difference between believing that a generalization will always work because they have seen a bunch of examples and developing an explanation for *why* it works in all cases. They were working on that idea in the math activity they just completed. Now, I wanted them to look at students' work to see when they are showing *why* the generalization holds.

Third, I wanted participants to look at students' ideas about how one can make generalizations for *all* numbers given the never-ending nature of numbers. How do the third graders in Dolores's case take it on? In Lucy's case, what is the difference between Amanda's position and Elizabeth's? Do they see that Elizabeth has taken on a sophisticated issue—one that Amanda may have already resolved for herself?

In their small groups, participants dug into the cases. In particular, they looked carefully at students' language in the first case, written by Dolores, to think about whether students had a generalization in mind. When students claimed that you could not make 33 using only even numbers and addition, were they saying that the sum of even numbers must always be even, or were they saying that they could not find even numbers that add to 33?

Kaneesha stated, "Mary, Ricky, D.C., Laurie, and Claudia all said that you cannot be sure because there are so many numbers out there. There are numbers they don't know about yet."

I asked, "What did we do that allowed us to make claims, even though we didn't try every number?"

Yanique said, "We had a visual model for *even*. We could look and see. If this representation shows an even number and it shows another even number, you see that when the amounts are combined, the total is even."

Denise added, "All of our arguments started from a definition. We do not see students in Dolores's class going back to their definition to understand what is happening. It seems like that is really important—to first clarify what the definition is so that everyone understands it. Then the arguments follow from the definition."

Yanique and Denise were making important points. I want to remember to refer back to these ideas when I can.

I also visited the group that included Phuong, Charlotte, and Madelyn. Phuong commented, "Some students are intimidated by the larger numbers or numbers they don't know. Eva is able to talk about even numbers that they know and what she believes about them, but is still uncertain about numbers she does not know."

Charlotte asked, "What does Eva need to believe in order to drop the 'numbers we know' and change it to 'all numbers'?"

Madelyn said, "I wonder whether other students were really thinking only about numbers they know, even if they do not say so."

That seemed to be a great question. I mentioned that this would be something she might investigate when she gets a chance to talk to her students for the homework assignment.

I chose to focus the whole-group discussion on Lucy's case about Amanda and Elizabeth. To start, I wanted to make sure everyone understood Amanda's argument, so I asked if someone would explain it.

Jorge started us off: "If you have two even numbers, you can count them both by 2s and land on the last number. So, if you put them together, once you finish one number you go right into the other and land on the last number again."

June said, "When I first read Amanda's argument, I was not fully convinced, but Madelyn had a great way of showing it. Madelyn's example sort of combines counting by 2s with Nadine's idea of coloring even and odd numbers. If you build an even number with cubes, you can go black, red, black, red—you start with black and every other cube (the numbers you say out loud when you count by 2) is red, and the last cube is red. When you color code the numbers you say out loud when you count by 2, you can really see it."

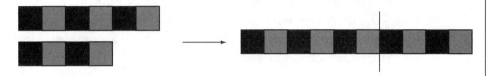

I pointed out that June was showing us two particular numbers, 6 and 4, so I asked if anyone would be willing to explain how this argument shows that it will work for all even numbers.

Risa said, "As Amanda said, you can count by 2s to any even number. So, if you build them black, red, black, red, any even number will end on red. Then, the next number will start again with black and will also end in red. When you put the two numbers together, you can see that every other number is red and it ends on red."

I asked the whole group, "Do you feel satisfied with Amanda's argument?"

Jorge said, "We already know that if you add two evens, the answer is even."

I explained, "I'm not asking whether you agree with her conclusion. I am asking whether you think she has come up with sufficient proof."

Madelyn said, "It seems to me she has offered a proof like the ones we have been using today. She has a way of showing what even numbers are—she says they are the numbers you get when you count by 2s—and her representation shows what happens when you add two even numbers."

Grace said, "Amanda's argument seems convincing to me." Several others said, "Yeah."

However, Phuong protested. "I am not ready to accept Amanda's argument." When I asked why, she said, "Because she arranges the cubes linearly, but it is so much clearer when the cubes are arranged in an array. If you arrange the cubes the way Amanda did, it is easy to miscount. However, if you show the pairs in an array, you just look at it and know the number is even."

I pointed out that the issue is not which proof we prefer, but whether Amanda's argument is valid. The premise of the claim is that we are starting with two even numbers. It is not a matter of whether someone might miscount, because the conjecture already states that the two numbers are even.

"Phuong," I asked, "if you accept this premise, that the two numbers are even, does Amanda's representation show that the sum is even?"

Phuong said, "I would not want to show that argument to a class because it is harder to see."

Rather than open her argument for discussion, I felt it was important for me to clarify what it means to assess the validity of an argument. "We are not thinking in terms of which proof we would most want to show a class. Instead, pretend that Amanda is in your class and has presented this argument. Even if you prefer a different argument, it is important to assess Amanda's reasoning in its own right. If her reasoning is faulty, it is important to help her find her error. However, if her reasoning is sound, it is important that it be acknowledged."

Phuong listened intently and nodded, as did others in the group. We went through the three criteria with participants explaining how Amanda's argument satisfied each.

Then I asked whether people understood Elizabeth's objection.

Grace said, "Well, one thing you can say about Elizabeth is that she is willing to try out someone else's idea."

I asked Grace for line numbers, and she responded, "Well, look at the whole thing starting at line 437. Elizabeth is working really hard to figure out what Amanda said."

June offered, "Elizabeth might accept the idea but needs more evidence."

Antonia said, "Elizabeth has not proven the idea for herself yet."

Phuong asked, "What do you think would give Elizabeth more confidence? Why do you think she does not have confidence to say the model will be right with all numbers?"

Vera said, "I was wondering the same thing. I am back to thinking about Eva in Dolores's case. She talks about 'numbers we know.' Why is Elizabeth not able to get beyond that?"

Charlotte said, "You know, I think Elizabeth is getting beyond the numbers she knows. It seems like Elizabeth knows that coming up with examples is not enough. It is a really big thing to say you can take *any* even number and add it to *any* even number and know you will get another even number."

Charlotte was saying something very important and I wanted the group to pay attention. I said, "Let's think about this for a minute. When we hear kindergarten or first-grade students make claims about all numbers, what are the numbers they are thinking about? Do they have a sense of numbers going on forever?"

Discovering Rules for Odds and Evens

I flipped back to Nadine's case and found the line I was looking for. "In Nadine's first-grade class, Jack says, at line 159, 'If you take two numbers that are even and put them together, you have an even number.' Now, we do not have a lot of time right now to dig into that case, but I just want to raise this as a question. It might be that Jack, as a six-year-old, can say that and be thinking about numbers up to 20 or up to 30 or up to 100. However, around third grade, students like Elizabeth begin to understand what it means for numbers to go on forever. So for Elizabeth, who understands that numbers extend way beyond what she can think of, it means something very different to make a claim about all numbers than it would be for a young child, perhaps Jack, who is thinking only about the numbers he has encountered."

Charlotte said, "Yeah, that is what I was thinking about. There is something important about Elizabeth's skepticism."

Risa said, "What about Amanda? Do you think she is like Jack, or is she saying something about all numbers?"

Yanique said, "It seems to me she is talking about all numbers."

I asked Yanique for a line number, and pointing us to line 468, she said, "After Elizabeth made her objection, Amanda was clear her argument would always hold. She said, 'Because 2s don't get to odds.' She knows if you keep counting by 2s, and you put one number on top of the other, you will never land on an odd number."

I said, "I think I would agree. My guess is that Amanda has worked through the idea Elizabeth is wrestling with. Amanda is presenting an argument that proves her claim about all even numbers, even though she cannot test all even numbers. Elizabeth has figured out something important, but she has more work to do to realize one can prove a claim about all even numbers.

"Of course," I continued, "we only have the evidence of the case and we cannot really know what Amanda was thinking. We do not have the opportunity to ask more questions of her. However, if something like this comes up with your students, you will have a chance to dig more deeply."

I looked at my watch and saw that time was almost up. "We have covered a lot of territory today. There is a lot for us all to think about. You will have an opportunity to work on some of these ideas as you complete your student-thinking assignment for homework."

I explained the homework assignment before distributing exit cards. As everyone had completed student-thinking assignments in previous seminars, I did not think we needed to spend much time on it. I did point out that, in addition to reading Chapter 2, they were being asked to write out the generalizations that students in each of the cases were working on.

Exit cards

As I said, that was a packed session, and I was pretty exhausted at the end. The exit cards affirmed that a lot of algebraic thinking had occurred during the session. The participants were thinking hard about the variety of issues we discussed. Several people wrote that they were surprised to discover that there is so much to learn about odd and even numbers. Risa said that before today she thought she knew all there was to know about odd and even numbers but that seeing the models with cubes gave her insights she had not seen before. 685

James was still thinking about the idea that Denise talked about in the session: Why are odds and evens based on 2 rather than other numbers? 690

Another idea that many people wrote about had to do with the development of students' thinking with regard to making generalizations. Phuong is still thinking that students are intimidated by large numbers and wonders how we can help them.

Yanique wrote, "Using the term *the numbers we know* actually makes me feel confident that students are aware that numbers are infinite yet makes me wonder how they can make generalizations with unknown numbers." 695

Lourdes wondered, "When do students move from 'specific number' instances to the concept of infinite numbers? When do they decide they can make a rule without testing every single instance? How does the concept of number structure develop in students?" 700

Denise commented, "The idea that older students can appear more uncertain because they are able to anticipate more possibilities is really significant because it can reflect a deeper level of thought than certainty implies."

Yanique wrote that her ideas about algebra have already changed. Everyone indicated that they found the session to be stimulating—some said fun. 705

Responding to the first homework

September 20

When I read the participants' homework that expressed their ideas about algebra, I did not learn too much more than what I had picked up in class. Though now I could hold onto who had positive experiences in algebra previously and who had not. 710

Although I usually respond individually to participants' writing, this time I thought it would be more effective to write to the whole group. I decided to use a letter to respond to issues that arose in their exit cards rather than those of their homework assignment, especially as there were some loose ends about odd and even numbers that might not be tied up in future seminar sessions. I wanted to 715

Discovering Rules for Odds and Evens

address those loose ends. In fact, I decided to send my letter out via e-mail so that people could read it and think about the ideas before we meet rather than wait until the next session.

September 19

Dear RAO seminar group,

720

I've read your preseminar writing and your exit cards, and I have got to tell you, based on our work together last Wednesday, as well as your writing, I am very excited about what is ahead of us in this seminar. It is a pleasure to be working with such a thoughtful and committed group.

725

Your writing and your ideas in the seminar session brought up lots of thoughts for me, particularly about odd and even numbers. Instead of waiting until I see you next—and by then, all of our thinking will be moving in a different direction—I've decided to share some of my thoughts with you now.

One question that was raised in the exit cards was why 2 is used to determine whether a number is odd or even. Instead of looking at whether a number can be split into groups of 2, why not see if it can be split into groups of 3 or some other number? The fact is, odd and even numbers are defined in terms of dividing by 2. We could think about similar questions with other numbers; there are just more categories to consider. For example, if you divide a whole number by 3 to see if it can be split into groups of 3, it might divide evenly without a remainder (let us call it type A) or it might divide with a remainder of 1 (type B) or it might divide with a remainder of 2 (type C). So you have three categories. Then you can ask similar questions: What happens if you add two numbers of Type A? Type B? Type C? What happens if you add a type A number with a type B number etc.?

730

735

You can ask the same kinds of questions about what happens when you divide a number by 4, giving you four categories of numbers. You can also pick any whole number to divide by and ask these questions. There is a name for thinking about numbers and operations in this way—modular arithmetic. It is just that we have special names—odd and even—for when you divide by 2.

740

Some of you were interested in what these ideas about odds and evens look like in algebraic notation. We can represent even numbers as 2x, where x is a whole number (or I could say 2y or 2n or 2b—I could use any letter) and odd numbers as 2x + 1 (and, again, I could use any letter to represent any whole number). The sum of two odd numbers looks like this:

745

$$2x + 1 + 2y + 1 = 2x + 2y + 2$$

750

(I used x and y to show that 2x + 1 and 2y + 1 might be two different odd numbers.) We could do more fiddling with the algebraic expression; but for now, can you see the image of the cubes in this equation? Do you see how the equation represents the two single cubes coming together to make a pair?

Many of you were also thinking about students, what it means for students to make a generalization and how students develop in terms of their understanding

755

of generalizations. Is it a strength or a lack of confidence when students realize that there are numbers beyond their conception that might behave differently than the numbers they know? How do students justify the generalizations they make? That is, what does proof look like for young students? How do these ideas connect to a typical algebra class?

These are all important questions that we will continue to ponder as we move into other mathematical topics. I imagine that you will be gathering information about some of these questions as you complete your student-thinking assignment for our next class. I'm looking forward to continuing our discussions then!

Best wishes,
Maxine

After I sent out the letter over e-mail, several participants wrote back to thank me. Charlotte was the only person who actually told me what she got from it. She wrote:

Thanks for sharing your thoughts after the seminar. It was quite a moment when I was able to look at the algebraic equation and actually have it make sense! I had two things in my head; I know that you can add in any order and I could see the 2 being 1 + 1 or vice versa as the pairing of the "leftovers." Looking at the equation with ideas in my head—or images of oddness and evenness—made it possible to actually think while I was looking at it.

Detailed Agenda

Introductions: Sharing ideas about algebra   (20 minutes)

Small groups (10 minutes)

Whole group (10 minutes)

The first activity of the seminar is designed to provide the participants an opportunity to introduce themselves by sharing their impressions of algebra. As individuals enter the room, organize them in cross-grade level groups and tell them to share ideas that they included in their preseminar writing. Also try to organize the groups so participants from different schools have the opportunity to interact.

Post the agenda for the session so participants will be able to see the whole plan. Opening the seminar in this fashion sends the message that the sessions will begin promptly at the designated time. Point out that 10 minutes have been set aside for this activity. Remind participants to give each member of the group time to share thoughts and ideas.

After 10 minutes, call the group together for introductions. In addition to stating their name, school, and grade level, you also want everyone to say a word or phrase that relates to his or her experience with algebra. Record the phrases on poster paper. Once everyone has spoken, comment on the list. You might want to highlight both similarities and differences that were expressed. Save this poster so you may display it in Session 8.

Definitions of even   (25 minutes)

Small groups (15 minutes)

Whole group (10 minutes)

In this activity participants will work with two different definitions of *even*:

- An *even* number is a number that can be split into two equal-sized whole-number groups, such as 6 (3 + 3), 16 (8 + 8), or 50 (25 + 25).
- An *even* number is a number that can be made up of pairs, such as 6 (2 + 2 + 2) or 16 (2 + 2 + 2 + 2 + 2 + 2 + 2 +2).

Participants will develop these definitions by examining a short scenario describing the thinking of two second graders. The object of the activity is for participants to clarify these two definitions and to form arguments to show that the two definitions describe the same set of numbers.

Display the prepared poster with the scenario about Hue and Julio (see p. 45). Organize participants in small groups to:

- Identify and write out the definition of even number that each student is using.
- Work on this question: "How do you know that numbers that are considered even according to one of these definitions would also be considered even by the other? How can you show that these definitions describe the same set of numbers; that is, how can you show these definitions are equivalent?"

Let participants know that cubes are available and that it will be useful for them to consider how to use the cubes to express their ideas.

As you circulate during the small-group work, you may need to clarify the definitions in response to participants' work. For instance, even though 15 can be split into two equal groups of $7\frac{1}{2}$, it is not an even number because it cannot be split into equal-sized whole-number groups. If questions about how the definition of *even* might apply to 0 or to integers, such as ⁻4 or ⁻10, let participants know that work with integers, including 0 and negative numbers, will be a focus later in the seminar. Suggest, at this point, that they consider these definitions only in terms of the set of positive numbers.

As participants offer models, ask them how the models apply to *all* even numbers and not just the specific numbers they built. If no participants built a model using cubes, make a suggestion to some of the small groups to try that approach. As you circulate during the small-group work, note the strategies that various participants use. These observations can help you plan for the whole-group discussion.

After 15 minutes, call the group together. Begin the whole-group discussion by collecting the definitions of even number. Once both definitions have been proposed and clarified, ask the participants to explain how they determined that the definitions describe the same set of numbers. If no group built a cube model, you should demonstrate the use of such a model to the whole group.

Then ask for a definition of an odd number. Possible definitions include:

- A whole number that cannot be arranged into pairs; that is, after making as many sets of two as possible, there would be one left over.
- A whole number that cannot be divided equally into two groups of the same size; that is, when split as equally as possible, one group will have one more than the other.

Both of these definitions are equivalent to "an odd number is a whole number that is not even." Each of these definitions establishes the fact that every whole number is either even or odd.

Proving the Equivalence of Two Definitions

Consider the following two definitions of *even* for counting numbers:

■ An even number is a number that can be split into two equal-sized whole-number groups.

■ An even number is a number that can be made up of pairs of whole numbers.

To prove that the two definitions are equivalent, one must show that

1. any number shown to be even by the first definition can also be shown to be even by the second definition
2. any number shown to be even by the second definition can also be shown to be even by the first definition

Consider: Based on the first definition, any even number can be represented by two stacks of cubes with the same number of cubes in each stack. (Look at the left diagram below.)

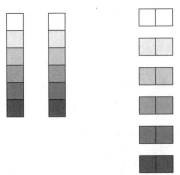

Now pair the cubes by matching the top cubes together, the second cubes together, the third cubes together, and so on. (See the diagram on the right above.) Since the original number can be represented as pairs of cubes, it is *even* according to the second definition.

Similarly, to show that any number that can be represented by pairs can also be arranged into two equal groups, first arrange an even number into pairs of cubes and then show, by stacking the pairs one on top of the other, you would have two stacks with the same number of cubes in each. See "Maxine's Journal" (pp. 16–17) for other arguments participants developed to demonstrate the equivalence of the two definitions.

DVD: Students talk about even numbers

(15 minutes)

Whole group

There are two different DVD segments for this discussion: one features first and second graders and one features seventh graders. In both segments the students are discussing what they know about even numbers. After viewing both segments, choose which is most appropriate for your seminar participants.

DVD segments of math classrooms provide practice for participants in listening to students describe their thinking in real time. The images from a classroom where students are sharing and building on each other's ideas can help bring to life the children in the print cases.

First and second graders talk about even numbers

Before showing the DVD segment, let participants know they will be discussing the following questions:

- How do the children describe odd and even numbers?
- How are the children using cubes as ways of showing their ideas?
- What questions would you want to ask the students and what would their response tell you?

Play the segment in two parts, stopping for discussion after a student, Jesse, points out the pairs in the two stacks. (This is approximately 4 minutes.) Ask participants for comments about what they have seen. One point that might arise is what meaning the students are attaching to the word, "even." Are they thinking of one of the definitions discussed in this session, or are they using the word "even" as a synonym for "level"? If this comes up, ask the participants what question they would pose to the class to clarify this issue. Then play the remaining section (approximately 1 minute) of the video. Discussion following the video should include their comments on the questions the teacher posed and what they learned from the student responses.

Seventh graders talk about even numbers

Before showing the DVD segment, let participants know they will be discussing the following questions and suggest that they take notes:

- What are the different ways the students identify how they know a number is even?
- What confusions about even numbers did the discussion surface?
- What questions would you want to ask the students and what would their response tell you?

After the DVD, ask participants to share the ways the students thought of even numbers (such as, the last digit is 0, 2, 4, 6, or 8; multiples of 2; and divisible by 2 with no remainder). Then ask what confusions about even numbers they noted. Choose some of the examples from the participants to ask what questions they would want to ask the students and what they would expect to learn about the students' thinking from their responses.

Math discussion: What is a mathematical argument? Adding odd numbers

(20 minutes)

Whole group

While the mathematical topic for this work is odd and even numbers, a major objective of this session is to help participants focus on what it means to produce

Discovering Rules for Odds and Evens

an argument that applies in general terms. The purpose is to use the arguments being developed by participants to introduce and explicate the three Criteria for Representation-Based Proof.

Pose the conjecture, the sum of two odd numbers is even. Have participants work in pairs to make an argument for how they know the conjecture is true. Some participants might begin by making a list of examples (4 + 3 = 7, 5 + 8 = 13, etc.). It is important to communicate that a list of examples is not sufficient to prove the generalization. Suggest the use of drawings or cubes to test how their arguments will apply to *all* pairs of odd numbers. Let participants know that after 5 minutes you will bring them together to share ideas and to develop some arguments as a whole group.

An important feature of making such arguments in the context of arithmetic is calling upon images or models of the operations. The argument for the conjecture considered here not only depends upon models of even and odd numbers but also on the action of putting things together as a way to represent addition. As participants work on the activities of this session, look for opportunities to ask questions about the models of the operations that they are using. Also ask questions to highlight how the argument is general; that is, how does the argument move beyond a set of examples and apply to a range of numbers? These questions provide the framework to introduce the Criteria for Representation-Based Proof into the whole-group discussion. (See "Introducing Proof" on page 12 and the Math discussion in Maxine's Journal on pages 19–22.)

If symbols are brought up, ask questions so that participants can discuss the links among the symbolic statements and the arguments expressed with the cube or diagram models. Read "Using Algebraic Notation" for more information. You should also acknowledge that some participants might be more familiar with symbols than others. Let the group know that while they will have opportunities to express ideas with symbols throughout the seminar, expressing arguments with words, cubes, and diagrams is valid and likely to be the methods that their students will use to express their ideas.

After one or two arguments have been shared and discussed, post the chart with the Criteria for Representation-Based Proof. Choose one of the participant's arguments and review it to illustrate how it satisfies the criteria. Note: You should display this poster at each session of the seminar as a reference.

Using Algebraic Notation

RAO begins with an exploration of odd and even numbers because many students notice and are interested in what happens when odd and even numbers are combined. Generalizations about the sums of odd and even numbers can be articulated clearly and convincingly proved with tools available to young children. These are the ideas to emphasize with RAO participants.

The statement of the generalizations and their proofs using algebraic notation requires a level of understanding that the RAO participants may not have. For this reason, we suggest that facilitators steer away from presenting algebraic notation in the whole-group setting. Notice that in "Maxine's Journal" (p. 23), the facilitator addresses issues of algebraic notation in small groups where participants are already using it and have some fluency, albeit with some errors. She also introduces the notation in written response to participants' homework and exit cards.

An even number is defined as a number that can be represented as $2n$, where n is an integer. (Integers are the counting numbers, their negative counter parts, and 0.) An odd number is one that can be represented as $2n + 1$, where n is an integer. Note that 0 is an even number and so are $^-2, ^-4, ^-6, ...;$ $^-1, ^-3, ^-5, ...$ are odd numbers.

To prove that the sum of two even numbers is even, consider two even numbers, $2n$ and $2m$, where n and m are integers. Note that we use two different variable names, n and m, to designate the two even numbers, $2n$ and $2m$. If we were to use just one name, $2n + 2n$, then n must take on the same value in both terms of the expression. For example, if $n = 3$, $2n + 2n = 6 + 6$; that is, when we write $2n + 2n$, we are not including all of the instances we need to consider. When we use two different variable names, they can take on different values or the same.

The sum of the two even numbers is represented by $2n + 2m$.

$2n + 2m = 2(n + m)$ by the Distributive Property.

Since $n + m$ is also an integer, $2(n + m)$ is an even number. Thus, the sum of two even numbers is the product of 2 and an integer, so it is also an even number.

To prove that the sum of two odd numbers is even, consider two odd numbers, $2n + 1$ and $2m + 1$, where n and m are integers.

$(2n + 1) + (2m + 1) = 2n + 2m + 2$ by applying the Commutative and Associative Properties.

$2n + 2m + 2 = 2(n + m + 1)$ by the Distributive Property.

Since $n + m + 1$ is an integer, $2(n + m + 1)$ is an even number.

Some participants may try to represent the same idea by using the letter "e" to stand for an even number and "e + 1" for an odd number. They may write, $(e + 1) + (e + 1) = e$ to represent the statement, "If you add two odd numbers, the result is an even number." In this instance the letter "e" is being used as a label for an even number but not as a variable. The fact that "e" is not being used as a variable can be seen by replacing "e" with a value, say 4. In that case, $(e + 1) + (e + 1) = e$ becomes $5 + 5 = 4$, a false statement. Sorting out the difference between using a letter as a *label* and as a *variable* can be confusing. For this reason, facilitators should discourage the use of such notation.

DVD: Adding odd and even numbers
(10 minutes)

Whole group

The second DVD segment shows fifth graders talking about adding three odd numbers. The DVD runs for about 5 minutes. Instruct participants to take notes focusing on how student thinking develops throughout the discussion. After watching the DVD, analyze the teacher's actions by asking, "Once one child explained that the sum of the three numbers is 27 and 27 is odd, why did Janelle continue the discussion?"

Break
(15 minutes)

Math activity: Operating with odd and even numbers
(40 minutes)

Small groups (20 minutes)

Whole group (20 minutes)

Distribute the Math activity: Operating with odd and even numbers, if you have not already done so. Because the whole-group discussion already addressed Question 1, participants should work in their small groups on Question 2, exploring what happens when two odd numbers are multiplied. Remind them to make arguments that use diagrams, cubes, and verbal expressions. Suggest that they consider how their argument satisfies each criterion for Representation-Based Proof.

After 20 minutes, call the group together for a whole-group discussion. As each argument is explained, provide time for participants to ask questions of each other. Lead them to explore the main themes:
- What definition of even or odd number is this argument using?
- How is multiplication represented in this argument?
- How does this argument extend to all whole numbers and not just to a set of specific examples?

Include arguments that use cubes and diagrams. If symbols are introduced, help the participants identify links between the symbolic argument and the pictorial ones. Also, remind the participants that they do not have to be comfortable with the symbolic forms at this point.

Case discussion: Students think about odd and even numbers
(30 minutes)

Groups of three (15 minutes)

Whole group (15 minutes)

Discussion of the cases in Chapter 1 should be related to these questions:
- How do students build mathematical arguments based on the structure of numbers, definitions, and the meaning of the operations?

- How does student thinking about making arguments develop across the grades?

Organize the participants into groups of three and distribute the Focus Questions for Chapter 1. Let them know they will have 15 minutes for small-group discussion. During this time, they should use the focus questions to examine students' differing ideas about what it means to prove a statement is true for all numbers, including whether it is even possible to do so.

Focus Questions 1 and 2 examine the thinking of third-grade students as they consider adding two even numbers. In discussing Focus Question 1, participants should note that the student comments range from "it isn't possible" to a restatement of what is to be proven. Focus Question 2 examines how one student offers a general argument and a second student struggles to accept the possibility that an argument can apply to all numbers.

Focus Questions 3 and 4 are based on Carl's case, Case 5, which presents the thinking of seventh-grade students as they work to define and then to use their definitions of even numbers.

Let participants know it is OK if they disagree with one another as they work on the focus questions. Also remind them to look for specific details in the cases to support their ideas. During the small-group work, keep participants focused on the cases by asking questions such as "Where in the case or at what line number is the evidence for your statement? What is it that these children think about making an argument that holds for all numbers? How do the students in this case express their generalizations or form their arguments? What do the students call upon to support their arguments?"

Begin the whole-group discussion by discussing Focus Question 2, which is based on Case 4. Have participants summarize Amanda's argument and Elizabeth's discomfort with the idea of being able to make a claim that would be true for all numbers. Finally, turn to Case 5 and discuss the thinking of Alex and Margot.

Homework and exit cards

(5 minutes)

Whole group

Distribute the "Portfolio Process" handout and answer any questions participants have about maintaining a portfolio. The portfolio can be a binder, a folder, or any other means of keeping a collection of their written work and your responses. Explain how important reading and responding to papers are for you, as the seminar facilitator; knowing their thinking helps you plan for the next session. They should also understand the portfolio assignments are carefully designed to move their ideas forward and will often be used as a basis for class discussion. Establish a routine for collecting and returning assignments.

You should also distribute the handout "If You Have to Miss a Session" or orally explain your expectations about attendance. Emphasize that the ideas of the seminar build from session to session; therefore, missing any session will jeopardize their continued understanding and active participation in class activities. Some facilitators suggest that participants who miss a session write an additional paper with their reaction to the cases from that chapter.

Distribute the "Second Homework" sheet and have participants read the portfolio assignment. Say, "Sometimes people read the cases in the chapter and wonder if their own students think about math ideas in similar ways. This assignment is a chance to check that out." Explain that these papers will be the basis for discussion with two other participants at the next session. Let them know this kind of assignment will be given three times over the course of the seminar.

Suggest they include their own analysis of the class or students in the paper. "Tell us what you make of what the students did. What did you learn about your students' thinking by looking closely at their work or writing? What questions does their thinking raise for you?" Finally, ask if there are any questions about the assignment.

As the first session ends, distribute index cards and pose these exit-card questions:

- What mathematical ideas did this session highlight for you?
- What was the session like for you as a learner?

Explain this procedure by announcing that each session will end with a period of reflective writing, known as "exit cards." This is an opportunity for participants to think about the experience of being in the seminar and sharing their feelings and learnings with you. Emphasize that their responses are invaluable as you plan future sessions. Exit-card questions are of two types; one is about the math content of the seminar and the other is about the seminar experience. Since participants are using the exit cards as a way to communicate with you, they should sign their names.

Before the next session ...

In preparation for the next session, read what participants wrote in their pre-seminar writing assignment in response to the questions, "What does algebra mean to you? When you hear the word *algebra*, what kinds of mathematical ideas come to your mind?" Then, choose quotes from the students' writings to include in a brief summary of their main points. Be sure each participant's ideas, if not their exact words, are represented in your response. You might also choose to respond to any issues or ideas that came up in the exit-card activity. For more information, see the section in "Maxine's Journal" (p. 32) on responding to the first homework. Make copies of both the papers and your response for your files before returning the work.

Session 1: Even and Odds

Clip One: First- and second-grade class with teacher Karen (5 minutes)

The students are sitting in a circle on the floor. The teacher, Karen, has a large tower of cubes. She breaks off some cubes so that she has two stacks of cubes that are the same height. Karen asks, "How can you tell this is an even number?" The students argue that the two towers are the same height and so the number has to be even.

After some conversation a student named Elizabeth shows that breaking 6 into 3 pairs is another way of showing that 6 is even.

The teacher connects the idea of finding pairs to the two towers, "Do you see pairs that go together?" Jesse shows how to look at the tower and see 2 and 2 and 2....

(Note: Stop for discussion.)

There is a break in the DVD and then the class returns. Now the teacher has two towers that are not the same height; one is taller by one cube. The class talks about whether the number of cubes in the tower is even or odd.

She takes off one more cube so that one tower is two cubes taller than another one. This time, there is disagreement about whether the number of cubes is odd or even.

Clip Two: Seventh-grade class with teacher Rich (6 minutes)

The segment begins with a brief narration indicating that it is early in the school year and that the class will be brainstorming what it is they know about even numbers.

The teacher poses two questions to the class: What is an even number? How do you know? He suggests they think individually for a while and then talk in their small groups.

The segment picks up again as Rich calls the whole class together to gather their ideas. The student ideas include:

> The last digit of an even number is 0, 2, 4, 6, or 8.
> Even numbers are divisible by 2 with no remainder.
> Even numbers are multiples of 2.
> Even means you can split it evenly, like 4 brownies and 4 people.
> Even numbers means you have an equal amount of stuff.

The teacher poses the question, "Is 876,547,289,634 an even number?" Some students say yes because it ends in 4. One student raises a question about the fact that it ends in 34 and suggests maybe it would need to end in 22 or 44 to be even. A student offers the example that 1,034 is even, and she knows because when she divides it by 2 the answer is 517 and 517 is a whole number.

Other students are not sure because 517 itself is not even. One student points out that 2 divided by 2 is 1 and 1 is odd, but they know 2 is even.

Clip Three: Fifth-grade class with teacher Janelle (5 minutes)

In this DVD clip, fifth-grade students are presenting arguments for how they know the sum of three odd numbers is odd.

First they look at $11 + 9 + 7$. One student says the answer is 27, so it is odd. The teacher asks the students what images they could use to explain this thinking. Brittany explains that she uses circles and pairs them. The teacher summarizes Brittany's method using the term *partner* as a way to represent even numbers.

Kim, a student, is thinking of the numbers as groups of people. She claims that with two of the odd numbers, the extra person from one number gets paired with the extra person from the second number. However, there is still an extra person in the third odd number who has no one to partner with. So, the sum of the three numbers will be odd.

Another student, Freddy, makes a different argument, indicating that the extra people from each of the three odd numbers makes 3. Then, when you try to pair the 3 leftovers up, you still have 1 left over.

The teacher restates both arguments.

Posters to prepare for Session 1

Poster 1:

Hue and Julio scenario

In a second-grade classroom the teacher commented there were an even number of children in the class that day.

Hue said, "I knew it was even because when we lined up for lunch everyone had a partner."

Julio said, "I knew it was even because when we split into two groups, the two groups were equal and no one was left over."

Poster 2:

Note: This poster should be displayed at every seminar session.

Criteria for Representation-Based Proof

1. The meaning of the operation(s) involved is represented in diagrams, manipulatives, or story contexts.
2. The representation can accommodate a class of instances (for example, all whole numbers).
3. The conclusion of the claim follows from the structure of the representation.

Math activity: Operating with odd and even numbers

As you work on these problems, make sure to use diagrams, cubes, and other visual representations (as the children did in Ingrid's Case 3) to make your arguments.

1. What happens when two odd numbers are added? Is this true for all pairs of odd numbers? How do you know? What kinds of arguments can you make to show your generalization is always true? What representations for odd and even and for addition do you call upon in each of your arguments? What definitions for odd and even numbers do your representations embody?

2. What happens if you multiply two odd numbers? Two even numbers? An odd and an even number? Is this true for all pairs of numbers? How do you know? What kinds of arguments can you make to show your generalization is always true? What representations for odd and even and for multiplication do you call upon in each of your arguments? What definitions for odds and evens are you using?

3. How do your generalizations and your arguments change if you consider addition or multiplication with three or more numbers? For example, what happens if you add three odd numbers? Four odd numbers? What representations for odd and even and for the operations do you call upon in each of your arguments? What definitions are you using?

Focus Questions: Chapter 1

1. In Case 1 in lines 36 through 45, Dolores reports how several students react when asked how they can be sure that adding two even numbers will always result in an even number. Explain the thinking of Ricky, D.C., Laurie, and Eva. How are Zoe and Mim thinking about this question?

2. In Case 4, lines 409 through 473, Lucy's students are discussing whether the sum of two even numbers is always even.
   - Explain Amanda's argument and her demonstration of her thinking.
   - What is Elizabeth's concern about Amanda's argument?
   - Do you think Amanda's argument applies to adding all pairs of even numbers? Explain how her argument works or explain how to modify it so it will work.

3. Describe the thinking of Zach and Brianna from lines 604 through 608 in Case 5. How is it the same or different from the thinking of the third graders in Case 1? What ideas does Carl bring to the class in his response in lines 610 through 619?

4. In line 629, Taylor raises a question about whether 4.4 is an even number. Then Carl challenges the class to create an argument for why they think 4.4 is even or not. Describe the thinking of Alex and Margot in this situation.

The Portfolio Process

As a participant in the DMI seminar Reasoning Algebraically About Operations, you will be expected to complete a writing and reading assignment for each session. For some sessions, you will also be asked to work on a mathematics assignment. In fact, you have already completed the first assignment when you wrote about your ideas about algebra and read the cases in the first chapter.

You will write a paper for each class session. Some of these writings will be read and considered in group discussion; all are opportunities to communicate with the facilitator of the seminar. The facilitator will collect your paper at the end of each session. During the following session, your paper will be returned to you with a written response from the facilitator. Please save both your writing and these responses in a folder that will serve as your seminar portfolio.

There will be a total of nine papers, one to be turned in at each session and one—the final portfolio review—to be completed at the end of the seminar. In all cases, the purpose of the assignments and written responses is to stimulate your thinking. The portfolio will be a record of your work and will also serve as a tool for reflection. Particularly toward the end of the seminar, you will find it very helpful to be able to look back over your work and think about how your ideas changed.

If You Have to Miss a Class

DMI seminars are highly interactive, providing many opportunities for you to express your own ideas and to listen to the ideas of your colleagues. Much of what you learn in the seminar is developed through small-group and whole-group discussions. You will get the most from the seminar, and will be able to contribute positively to the learning of everyone in the group, if you make every effort to prepare for each session and attend the entire time.

Frequently, ideas that are introduced in one session are expanded upon and developed more fully in later sessions. Thus, every session is important. If you find that you are unable to attend a particular session, or might miss a part of a seminar (by coming late or leaving early), please contact the facilitator as soon as possible. Make arrangements to turn in assignments and to obtain copies of the assignments that you will miss.

When you are absent, you are also responsible for turning in a reflective paper describing your reaction to the cases that were discussed at the session you missed. This is in addition to the regularly assigned work. You also need to turn in your work on the math activity that was completed while you were absent. If at all possible, meet with a classmate to discuss the session you were not able to attend.

These requirements are designed to insure that a missed session will minimize any loss of learning.

Second Homework

Reading assignment: Casebook Chapter 2

Read Chapter 2 in the Casebook, "Finding Relationships in Addition and Subtraction," including both the introductory text and Cases 6–10.

As you read, note the generalizations that are present in the students' work. Write out the generalizations in words and/or symbols and bring them to the next session. Your work will be shared with other participants and then turned in to the seminar facilitator.

Writing assignment: Examples of student thinking

It is likely that reading the cases and working on the math in this seminar have made you curious about how your own students think about odd and even numbers. This assignment asks you to examine the thinking of your students.

Pose a question to your students related to odd and even numbers. You might ask a question taken directly from one of the cases in Chapter 1. Then think about the students' responses. What did you expect? Were you surprised? What did you learn? Write up your question, how your students responded, and your reactions to their responses. Include specific examples of student work or dialogue. Examining the work of just a few students in depth is very helpful.

At our next session, you will have the opportunity to share this writing with other colleagues. Please bring three copies of your writing to turn in.

Note: You will be asked to prepare similar assignments that involve investigating students' thinking for Session 4 and Session 6. Check your classroom schedules and lesson plans to be sure you will be able to complete these assignments.

REASONING ALGEBRAICALLY ABOUT OPERATIONS

Finding Relationships in Addition and Subtraction

Mathematical themes:

■ What are the generalizations that underlie various computational strategies for addition and subtraction?

■ How do we draw on models for representing addition and subtraction, such as visual images, story contexts, and number lines, to express and justify generalizations?

Session Agenda

Sharing student-thinking assignment	Groups of three	35 minutes
Discussion: Articulating Lola's students' generalizations	Whole group	20 minutes
DVD: Subtracting 1 from an addend	Whole group	20 minutes
Math discussion: Proving generalizations behind our computational procedures	Whole group	30 minutes
Break		15 minutes
Case discussion: Students' generalizations and teachers' moves	Small groups Whole group	30 minutes 25 minutes
Homework and exit cards	Whole group	5 minutes

Background Preparation

Read

■ the Casebook: Chapter 2

■ "Maxine's Journal" for Session 2

■ the agenda for Session 2

■ the Casebook, Chapter 8: Section 4

Work through

■ the Focus Questions for Session 2 (p. 86)

■ the "Optional Problem Sheet 1" (p. 87)

Preview

■ the DVD segment for Session 2

Poster

■ Criteria for Representation-Based Proof

Materials

Duplicate

■ "Focus Questions: Chapter 2" (p. 86)

■ "Optional Problem Sheet 1" (if appropriate) (p. 87)

■ "Third Homework" (p. 88)

Obtain

■ DVD player

■ index cards

■ cubes

Prepare a poster

■ Student names from DVD segment (p. 85)

Expressing Generalizations

Throughout the RAO seminar, participants are asked to express generalizations that are implied in the work of students in the print and DVD cases. Writing the generalizations in words is not a trivial activity. It requires a preciseness of language and demands careful attention to students' thinking. Expressing a mathematical idea in careful language is an important part of the work of this session and the ones that follow.

In addition, participants may also work to express the same generalizations in symbolic form. Because they will already have in mind the idea they are working to express, this exercise helps them attach meaning to algebraic notation.

There is often more than one way to express a given generalization both in words and with symbols. For example, consider the statements of Amalia, Manuel, and Coleman in the DVD as they are summarized below:

Amalia:	If $10 + 9 = 19$, then $9 + 9 = 18$.
Manuel:	If $10 + 10 = 20$, then $9 + 9 = 18$.
Coleman:	If $9 + 9 = 18$, then $9 + 8 = 17$.

When expressing a generalization that underlies these examples, participants might say

- If you add two numbers, you get a particular sum. If you reduce one of the numbers by 1 and then add, the sum will be 1 less.
- If you subtract 1 from both addends in an addition problem, the sum will be 2 less.
- If you subtract any number from one of the addends, then you subtract the same amount from the sum.
- If you subtract an amount from one or both of the addends, then you subtract the total of that amount from the sum.

The variation in these statements is related to the degree of generality in expressing the idea. The first two statements follow quite explicitly from the examples offered by students. The latter two statements are more general and include cases in which the amount subtracted is not 1.

When expressing the first idea in symbolic notation, participants might offer the following:

If $a + b = c$, then $a + (b - 1) = c - 1$. In this version, c is used to represent the sum $(a + b)$.

Written more compactly, this same idea can be expressed as $a + (b - 1) = (a + b) - 1$.

For participants who are new to using symbolic notation meaningfully, the compactness of the second version of the generalization and the fact that $(a + b)$ represents both the instruction to add a to b and also the result of that addition requires significant work to understand. These participants may be more comfortable with the first version in which it is stated explicitly that c is the sum of a and b.

When symbolic expressions are introduced into the seminar, allow time for participants to test out the symbols with particular values. For instance, suggest that they let $a = 5$ and $b = 7$ and write out the resulting numeric expressions.

Below are possible ways to express the generalizations from the print cases in Chapter 2:

Case 6: If one number is larger than another number and the same quantity is added to both, the sum of the larger number and the added quantity is larger than the sum of the smaller number plus the same quantity.

If $a > b$, then $a + c > b + c$.

Case 7: If 1 is added to one addend, the sum increases by 1 as well.

If $a + b = c$, then $a + (b + 1) = c + 1$.
Or $a + (b + 1) = (a + b) + 1$

Cases 8 and 10: If an amount is added to one addend and that same amount is subtracted from the other addend, the sum remains the same.

$a + b = (a + x) + (b - x)$
Or $a + (x + c) = (a + x) + c$ (In this example, the quantity b is seen as separated into two parts, x and c.)

Case 9: If the number being subtracted is reduced by a certain amount, the difference will be greater by that same amount.

If $a - b = c$, then $a - (b - x) = c + x$.
Or $a - (b - x) = (a - b) + x$

Maxine's Journal

October 1

This was another packed session. Again, I wanted to work on the idea of generalizations: what they are, how students express them, and how to use representations as a convincing argument that the generalizations hold for *all* numbers. In this session, we were working on generalizations about whole-number addition and subtraction. | 5

Sharing student-thinking assignment

I arranged participants in groups of three to share their student-thinking assignment. They began by reading the other two participants' assignments and then discussed the set. In a couple of the groups, the participants talked about how hard it is to record students' ideas. I shared with them some of the strategies I have heard teachers describe in the past—using a tape recorder, | 10 asking students to slow down so you can write down their words as they're speaking, or jotting down just a couple of notes during class and writing the idea out after the class ends. "The thing is, no matter what method teachers use, they often say it changes what students say—they become more thoughtful. Sometimes, even when the teacher is not preparing for a student-thinking | 15 assignment, a student will say, 'I am about to say something important, so you should turn on the tape recorder,' or 'Don't you want to get a pen and paper to write down what I am saying?'"

Some teachers reported they had difficulty getting their students to take on the task that had been assigned to them. The students did not understand why | 20 they needed to do anything more than find answers to addition problems. These students will need to learn how to think differently in math class.

Some of the teachers seemed to have successfully recorded students' discussions and enjoyed the opportunity to compare what they had learned regarding their students' thinking about odd and even numbers. I am looking forward to | 25 reading these assignments and writing a response to each participant.

Discussion: Articulating Lola's students' generalizations

To start the whole-group discussion, I wanted to be clear about the transition from last week's work and their student-thinking assignments to the work we were about to begin. "From what I heard, your papers detailing your students' thinking about odds and evens sound very interesting. I am looking forward | 30 to reading what you wrote. Now, we are going to move on to something else.

As I explained last week, one of the main things we will be looking at in this course covers the generalizations that arise in students' work with numbers and operations. Last week, we thought quite a bit about odds and evens. We started with odds and evens because it's a context in which young students often make generalizations. In this second session, we will be looking at the generalizations embedded in the computation work of elementary- and middle-grade students. What generalizations are implicit in the computation moves students make? What does it take to make those explicit?"

I reminded the group that they had written out, in common language and/or algebraic notation, the generalizations the students were engaged with in each of the cases of Chapter 2. I pointed out that in these cases, students were operating with a generalization in mind, sometimes articulated, sometimes not. Our exercise is to express the generalization that students seem to be acting on. To start, we would look at Lola's case of kindergartners playing Double Compare. "Let's look at the generalizations you wrote out in English." (See Casebook page 31 for an explanation of this game.)

As participants offered their statements, I wrote them on the board.

Antonia said, "If two cards are the same and two cards are different, then the one with the larger of the different numbers says 'me.'"

I asked for comments about Antonia's generalization, and Madelyn said that she was not sure about it. "I assume the statement refers to the context of the card game, but still—I do not know if this is too picky or not—but what if one person has the two numbers that are the same? Like what if one person has 5 and 5, and the other person has 6 and 2. It satisfies the conditions, but it is wrong. The person who has 6 doesn't win."

M'Leah suggested rewording Antonia's statement. "If each person has two cards, and one of the cards each person has is the same, then the person with the higher different card says 'me.'" I looked around for comments on this version. Madelyn said this wording worked for her, and others nodded.

Mishal said she wrote it out algebraically. I responded, "We will get to that in a moment. First, I want to hear what people wrote out in English."

Lorraine gave us her version. "A number plus a big number is more than a number plus a small number." We talked about some of the assumptions in this statement. The two times she says *a number*, it means "the same number." When she says *a big number* and *a small number*, she means "in relation to each other." For example, if you have the cards 5, 3 and 5, 1, then 3 is the big number, even though it is not the biggest number in the two hands. In this case, it is called the big number because it is being compared to 1. However, if you have the cards 2, 3 and 2, 6, then 3 is the small number, even though it is not the smallest number in the two hands. Here, 3 is small because it is being compared to 6.

When I taught this seminar previously, I asked the participants to rework their generalizations to eliminate the ambiguity. It was interesting to see how difficult that was, and it highlighted the power of algebraic notation. The statement for Lola's case that the group liked best was, "When you compare two addition expressions, each with two addends, if they have an addend in common, compare the other addends. The expression with the greater addend has the greater sum."

However, this time, I wanted to focus on the difference between the two statements that had been offered. We had the following two statements on the board:

- If each person has two cards and one of the cards each person has is the same, then the person with the higher different card says "me."
- A number plus a big number is more than a number plus a small number.

I said, "Let's look at these two versions of the same idea. We all agree they both describe the generalization behind the students' actions in Lola's case. What is different about the two statements?"

Phuong pointed out that one is about numbers and the other is about cards.

This seemed very significant to me. "Yes, that is a big difference. We do not have much evidence about whether the students are thinking only in terms of the card game or if they are thinking about numbers more generally. This is a very important issue to notice and to keep in mind when working with students. It might seem that they have a general idea about numbers, but it might turn out that when confronted with the same mathematical idea in a different context, they will not recognize it. Their generalization might not extend beyond the card game." I paused to see if there were comments about this idea.

When nobody spoke, I asked if they saw any other differences between the two statements. Again, nobody commented, so I asked if the participants understood what I was talking about. M'Leah said yes, because it is something she works on with her first graders. "We play all kinds of games in which students have experience with addition combinations. I point out that when we play one game, they find 4 cubes and 3 cubes make 7 cubes. Does that mean when we play another game, they'll find 4 chips and 3 chips make 7 chips? Does it mean 4 students and 3 students make 7 students? Are they thinking just about the particular context they are in, or are they finding something that pertains to other contexts, too?"

Denise said that the same issues come up in the upper grades, too. You work on an idea in one context, and you think the idea is really solid for the students, but when you change the context, you realize the idea has disappeared.

I pointed out that it is not exactly that the idea is gone, but that the idea they had in the first place was not as general as you thought it was. That idea, confined to the original context, is likely to still be there. It's something you can draw upon as you work to help them understand the more general idea.

There was still another difference between the two statements that I wanted to highlight. I underlined the word *plus* in the second statement. "Do you see that the second statement is about addition, but the first statement does not say anything about an operation? This difference might seem pretty subtle, but I think it might be pretty important. As we explore Chapter 3, we will see that sometimes we think students are talking about addition, but then students reveal that they are thinking about numbers and not a particular operation. Addition can be tricky because sometimes you can hardly tell there is an action. So, when the students are playing this game with cards and trying to decide which pair has more, do they have the idea of addition in mind? We will be thinking about this issue more at our next session."

Phuong said, "What these students are doing reminds me of my algebra experience. If two things are the same, you can cross them out. Kindergartners have that figured out."

Phuong's statement was a sloppy articulation of an algebraic strategy most of us had learned, but we understood what she meant. A bunch of people laughed at the idea of kindergartners understanding a concept that they worked on in algebra. Of course, I had just been working on the point that we must not attribute to students generalizations they have not yet made. However, the idea students were working with *is* related to Phuong's idea. With this segue, I suggested we return to what Mishal was ready to show us earlier. "She said she represented kindergartners' generalization in algebraic notation."

However, before we could hear from Mishal, Grace presented her version of symbolic notation:

$$a + (>b) > a + (<b)$$

As she explained to me what to write, a number of people began to comment that it was incorrect. Before I could talk about Grace's notation, it was important to point out that we need to be more respectful as we respond to what someone has offered. Then I asked Grace about what she wrote. She said, "I was not really sure how to write it, but I decided to take a stab at it. If you add to a some number larger than b, the sum will be more than if you add to a some number smaller than b. That is what I was trying to write."

I said that I could see that interpretation, but that what she had written is not conventional notation. I explained, "Even though I can see what you mean, mathematicians do not write '$>b$' to mean some number greater than b. They might write '$x > b$,' which means that x is some number greater than b.

Then I turned back to Mishal, who offered two ways to state the generalization:

If $a > b$, then $a + c > b + c$.
If $a > b$, then $c + a > c + b$.

Risa said that the second line looks just like the second of our written statements: "A number plus a big number is more than a number plus a small number."

Denise pointed out that it is much easier to read the algebraic notation. "You do not have to go through all the talk we did about what 'a number' and 'big number' and 'small number' mean."

Leeann said it is easier to read the algebraic notation if you understand what it is saying.

I suggested that, for people who do not already know how to read algebraic notation, this might be a way to start making sense of the symbols. They already understand the written statement of the generalization, so if Mishal's version in algebraic notation says the same thing, where do c, a, and b show up in the written statement?

I suggested we look at the algebraic notation with particular numbers in mind. What if we had these cards: 4, 2 and 4, 3? What are the values of a, b, and c?

June answered for us: c is 4, b is 2, a is 3.

What about 5, 6 and 4, 6? Again, there was no problem. I checked in with Leeann who identified a as 5, b as 4, and c as 6.

What about 3, 3 and 5, 3? At this point, there was some confusion—there were too many 3s! Denise pointed out that this didn't matter; a is 5, and b and c can both be 3. I had the group refer to our written statements to see if they covered this example. Once the group was OK with the written statement, we went back to algebraic notation. I explained that when c appears twice, it must refer to the same number in both places. If we use different letters, the numbers they represent might or might not be different.

Finally, I asked about 4, 2 and 6, 3. Everyone was pretty clear that this example was not covered by the generalization we were working with because all four numbers were different. Madelyn pointed out that this kind of example was discussed in Lola's case. "The students might say that 6 and 3 is more than 4 and 2 because 6 is more than 4 and 3 is more than 2." Mishal said that she had written out that generalization, too: If $a > b$ and $c > d$, then $a + c > b + d$.

DVD: Subtracting 1 from an addend

Next, we looked at a DVD clip of a first-grade class in which the teacher asks for the answer to 9 + 9 and then 9 + 8. It seems that some of the students are acting on a generalization they have in mind, though they do not state it as a generalization. "As you watch the DVD, I would like you to take notes on the method that each child uses to compute either 9 + 9 or 9 + 8. In particular, I want you focus on what Carlos, Amalia, Manuel, and Coleman do."

Finding Relationships in Addition and Subtraction

After the DVD segment, some of the upper-grade teachers commented on the fidgeting of the students in that class. M'Leah laughed, "They're first graders! That's how first graders are!" Then we got to work, identifying how different students determined that $9 + 9 = 18$ or $9 + 8 = 17$. The participants noted the following statements that were made by the students in the DVD segment:

CARLOS: I take 9 in my head and count up.

AMALIA: $10 + 9 = 19$, so $9 + 9$ is 1 less.

MANUEL: $10 + 10 = 20$. So for $9 + 9$, I count back 1 and another 1 to get 18.

COLEMAN: Because $9 + 9 = 18$, then $9 + 8$ is 1 less.

I asked participants to work in small groups to state the generalizations behind Amalia's, Manuel's, and Coleman's strategies. Some participants are still finding it difficult to express the generalization. The clearest written statement was, "When you add two numbers, if you subtract 1 from one addend, you subtract 1 from the total." Again, several people wanted to write it algebraically: If $a + b = c$, then $a + (b - 1) = c - 1$.

In my own head, I thought $a + (b - 1) = (a + b) - 1$. When I shared that with Risa during the small-group discussion, it seemed to be too terse for her. I think the issue is that it is hard for people to see "$a + b$" as representing both the addition and the total at the same time. By writing the two equations, they can keep those ideas separate. This is something I will watch out for as the seminar continues.

During the whole-group discussion, we looked at the two versions of the generalization, in written and in algebraic notation (not my version, but the version participants wrote).

Math discussion: Proving generalizations behind our computational procedures

I pointed out to the group that, so far, we have been identifying the generalizations behind the actions of kindergartners and first graders as they begin to compute. "Now let's think about the generalizations underneath some of our own ways of doing things. Let's start with some mental math."

I presented the problem $39 + 18$. I chose those numbers specifically to elicit a particular strategy, and Vera immediately complied. After we determined that the total is 57, Vera shared her strategy: $40 + 17 = 57$.

"And how did you get $40 + 17$?" I asked.

"I took one from the 18 and gave it to the 39."

I wrote, "$39 + 18 = 40 + 17 = 57$," and asked if people were OK with the equal sign being used that way. I was checking to make sure the teachers do not think of "=" as meaning "the answer is the next number." Because everyone nodded, I went on.

I asked if Vera's method will always work, no matter what two numbers you add: take 1 from one number and give it to the other. Everyone said yes. Denise pointed out that you do not always want to do that; it depends on the numbers you start with. For example, if you are adding 36 + 15, it is not particularly helpful to change it to 37 + 14. I acknowledged that it might not be helpful for the mental arithmetic, but it still is true that 36 + 15 = 37 + 14. Then when I asked how they *knew* this rule always worked, they seemed baffled. What was I asking? They already explained that you take 1 from one number and give it to the other.

"OK," I said. "Let's look at 39 − 18. We take 1 from one number and give it to the other. Now, we have 40 − 17 = 23. Is 23 the answer to the original problem?"

Jorge said, "No, 39 − 18 = 21. You can't do that."

I said, "But everyone told me it always works, to take 1 from one number and give it to the other. Here, we see it works sometimes and not other times."

Kaneesha said, "It works for addition but not for subtraction."

I then asked the group to write out our generalization. It took some effort, but we came up with the following: "If you are adding two numbers, you can subtract 1 from one addend and add 1 to the other, and the total stays the same." Algebraically, the statement is shown as $a + b = (a + 1) + (b - 1)$.

From here I could define their task for the small-group work. "In our last session, we used representations to prove rules for operating with odd and even numbers. Those representations needed to show the operation and needed to be flexible enough to accommodate any whole number. To justify your rule for adding numbers, I want you to come up with a representation that shows addition and proves that if you take 1 from one addend and give it to the other, the total will stay the same."

I put up the same poster I had in Session 1 to remind participants of how we are thinking about proof.

Criteria for Representation-Based Proof

1. The meaning of the operation(s) involved is represented in diagrams, manipulatives, or story contexts.

2. Representations accommodate a class of instances (for example, all whole numbers).

3. The conclusion follows from the structure of the representation.

The participants needed only a few minutes to come up with representations to share with the whole group.

Finding Relationships in Addition and Subtraction

Kaneesha held up two towers of cubes.

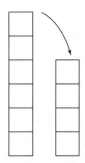

She explained, "It doesn't matter how many cubes are in these towers. Addition gives the amount altogether. You can move a cube from one tower to the other, and the total stays the same. Actually, you can move any number of cubes from one tower to the other, and the total stays the same. The process is not just taking 1 from one number, it's taking any amount from one number."

I asked Kaneesha if she wanted to revise our generalization. It now became: "If you are adding two numbers, you can subtract any amount from one addend and add that same amount to the other, and the total stays the same." In algebraic notation, the generalization was now, $a + b = (a + x) + (b - x)$.

Phuong said she pictured a number line with a bead on it. The number line started at 0, and we see $18 + 39$ by going to 18 and then moving 39 beyond that.

Phuong explained, "Right now the bead is at 18, but it can move left to 0 or right to 57. If it moves to the left, you are taking some amount off the 18 and adding it on to the 39, but 57 stays where it is. If you move the bead to the right, you add some amount onto 18 and subtract that same amount from 39."

James said that his model was pretty much the same as Phuong's, but he was thinking of a rectangular chamber with a barrier in it.

The chamber represents 57 altogether, with 18 on one side of the barrier and 39 on the other. You can move the barrier either way. Whatever is added to or subtracted from the 18 is subtracted from or added to the 39.

Madelyn said, "I was thinking about a story context. Let's say I was out picking apples with my son. At the end, we each had a bucket of apples, but his was too heavy for him to carry back. So I put some of his apples in my bucket. We still have the same number of apples between us."

These were all useful representations for justifying our generalization. In fact, as people had worked on each of their representations, they all extended the generalization from taking 1 from one addend to taking *any amount* from one addend.

At one point, Risa said, "It's like what my mother used to say, borrowing from Peter to pay Paul."

Was this one of those metaphors again—like two negatives make a positive from the last session—that have some kind of resemblance, but do not really explain anything? I asked Risa how it was like that, and she said, "Well, if you don't have anything to give Paul, you have to take that same amount from Peter. I guess it's not really the same because my mother always said not to do it."

It did not seem like a good use of time to consider if that saying actually fits the mathematics under discussion, so I moved on.

Now I wanted the group to think about subtraction. I reminded them that when we checked to see if this addition generalization applies to subtraction, we found that it does not work. "Starting with 39 – 18, if you take 1 from 18, and give it to the 39, the problem is different. If you take 1 from 18, what do you do with the 39 to make an equivalent subtraction problem?" I wanted them to work on that problem in small groups, find the generalization that works for subtraction, and come up with a representation to show why it works.

When we came back together as a whole group, participants offered the rule for subtraction: "Whatever you do to one number, you do to the other." I asked, "Like double it? If you double both numbers, do you get the same answer?" Some people needed a moment to actually double the numbers and see what happens. They soon clarified that whatever you add to or subtract from one number, you do the same to the other. It took a while to get the algebraic representation, but finally we got to $a - b = (a + x) - (b + x)$ and $a - b = (a - x) - (b - x)$.

Antonia offered a representation with cubes. She held up two towers and said that when you subtract, the difference between the heights is the answer.

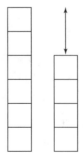

"I can take cubes off of the bottom or add cubes onto the bottom. As long as I add or subtract the same amount for both towers, that difference is going to stay the same."

M'Leah said she used a number line.

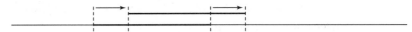

"I've got these two numbers, and the distance between them is what you get when you subtract. Now, let's say I put something rigid between the end-points, that's the difference. I can move it left and right, and when I do, the endpoints move the same amount." 335

Grace said that when she looked at a number line, it was clear what was going on when we took 1 from the 18 and added 1 to the 39. "It's so clear that when you change the numbers that way, you will have a bigger answer." 340

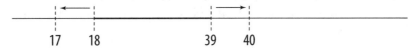

Madelyn said that she used a number line, too, but hers looked completely different. Her number line looked like this to illustrate 53 – 14 and 49 – 10.

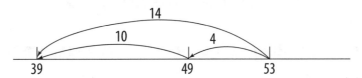

Madelyn explained, "I decided to put numbers on it, but I think it works more generally. To show 53 – 14, I can start at 53 and move 14 to the left to get 39. If I subtract 4 from 53 and also subtract 4 from 14, then I have 49 – 10, which is also 39." 345

Jorge said, "That is not how you subtract on a number line!" M'Leah and Grace also indicated that they could not make sense of what Madelyn had done. 350

It is important for participants to be familiar with two different ways of representing subtraction on the number line, and so I brought their attention to this issue. I said, "Remember in Making Meaning for Operations, in the session on whole-number subtraction we considered lots of different word problems that were all modeled by 7 – 2 = 5. I think this issue is related. Most of 355 the representations you presented today had to do with finding the difference between two amounts. Antonia held up two towers and showed how subtraction indicates how many more cubes one tower has than the other. Looking at the number line, we can mark the positions of the two numbers and find the distance between them." 360

"But we can also think of subtracting as removing some amount. Take one of Antonia's towers and take some cubes away. Subtraction gives you the number of cubes that are left. That is what Madelyn is showing on the number line. Start at 53 and move back 14—that is like removing 14. It lands you on 39: 53 – 14 = 39. Her picture also shows 53 – 4 = 49 and 49 – 365 10 = 39."

Charlotte said, "I can see that Madelyn's number line shows subtraction, and I can see that it shows 53 − 14 = 49 − 10. However, I don't see how it proves our generalization."

I suggested we label the number line with variables to see what it would look like.

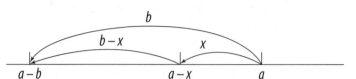

When the number line was labeled this way, Denise nodded and muttered to herself, "That's very nice." However, I was not sure where everyone stood.

I told participants they had been doing a lot of good, hard thinking. I suggested they might want to copy Madelyn's representation so they could go back to it later. Then they should take a break.

Case discussion: Students' generalizations and teachers' moves
Small groups

With all the work we had done to state generalizations in the first part of the session, people were eager to share and revise the generalizations they had written out for homework. They recognized that the generalization the students in Kate's class were working on was the same as the one we had just discussed—"If you are adding two numbers, you can subtract any amount from one addend and add that same amount to the other, and the total stays the same." Similarly, the generalization from Maureen's case was like the one we derived from the DVD segment. The groups I visited were now using the algebraic notation: If $a + b = c$, then $a + (b + 1) = c + 1$. I did not see anyone who had written $a + (b + 1) = (a + b) + 1$ for homework, but now Madelyn asked if that would be a correct way to represent the idea.

It seemed that a few people had misunderstood what was happening in Monica's case. Apparently, when they saw that students incorrectly calculated 145 − 98, they assumed that students had made the error they see frequently which some educators call "subtracting up." Students might say, "Because you can't subtract 8 from 5, just turn it around: 8 − 5 = 3. Then 14 − 9 = 5." Lourdes had made this assumption, but when I asked her to finish the calculation the way these students would, she realized the result would be 53 instead of 43. I told her, "The students in this case were actually thinking something different here to come up with an answer of 43. What was that?"

Mishal pointed out that they were comparing 145 − 98 to 145 − 100. "So the answer to 145 − 98 has to be 2 away from 45. They figured that because 98 is 2 smaller than 100, the answer should be 2 smaller than 45."

Finding Relationships in Addition and Subtraction

Grace pointed out that by the end of the lesson, Rebecca and Max had stated the generalization clearly, "If you take away smaller, you end up with bigger." and "The less you subtract, the more you end up with and the thing you end up with is exactly as much larger as the amount less that you subtracted."

Even though they could identify the generalization from Monica's case, it seemed that quite a few people were still struggling to figure out what it meant.

Kaneesha said, "I could look at that picture, but I wasn't sure I got what it meant at first. It really took Rebecca's comment about the bread to help me see it. If you have a hunk of bread and take a small bite, you end up with more left than if you take a large bite."

Jorge agreed that it was Rebecca's image that helped him. "When she puts it like that," Jorge said, "it's so clear what's happening with the numbers. Whenever I see a subtraction problem, I can think about breaking off a hunk of bread. If I take less, I leave more. Wow."

Risa said that it was easier for her to see on the number line rather than on the blob that Brian drew in Monica's class.

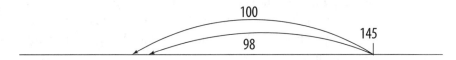

Risa explained, "If you start at 145 and go back 100, you end up farther to the left than if you go back 98."

I noticed that Risa represented subtraction on the number line as Madelyn had. Charlotte, who was looking on, said that it is easier for her to see subtraction as a difference.

Charlotte said, "When you look at the number line, you can see the distance between 98 and 145 is greater than the distance between 100 and 145. So 145 − 98 is more than 145 − 100."

I asked both Risa and Charlotte if they could change their representations a bit to illustrate the generalization rather than sticking with 98, 100, and 145. Risa said, "Sure, all I have to do is erase the numbers. If I start at one place, if I jump back more, I end up more to the left than if I jump back less."

Charlotte looked at her number line for a while, and then said it works for hers, too. "If you start at a smaller number, you have to go farther to reach your destination than if you start at a larger number."

At Denise's table, I saw that they had written out the generalization in algebraic notation: If $a − b = c$, then $a − (b − x) = c + x$. I told them this was correct, but suggested that they could try to write it as one equation instead of thinking of it as an if-then statement. Denise then wrote out, $a - (b - x) = (a - b) + x$.

Denise then paused, wrote $a - (b - x) = a - b - (^-x)$, and said, "When I took algebra in high school, I learned to distribute the subtraction sign. We also learned that $- (^-x) = x$. So you get this equation: $a - (b - x) = a - b - (^-x) = a - b + x$." She then wrote out $a - (b - x) = (a - b) + x$ and said "That's how you get to this. The thing is, now I am looking at that same equation with a completely different sense of what it means. I don't have to think about distributing the subtraction sign, and I don't need to think about $- (^-x)$. It feels really different to think, the less you subtract, the greater the result. *That's* what the equation says."

I pointed out to Denise that this is what the equation says as long as x is positive. When she nodded, I mentioned that we would be getting into negative numbers in the next few sessions.

Whole group

By the time we moved to the whole-group discussion, my sense was that the group was finished looking at those generalizations. I decided to move right to the question of what the teacher was doing to open up these discussions in the classroom.

Risa said that she was interested in what Kate was doing in Case 8 around line 330. In this case, students were working on the idea that when adding two numbers, you can take some from one addend and give it to the other and get the same answer. Risa said, "They started with the manipulatives and then other numbers, and then they expanded on their ideas."

Kate's case is a good one for looking at a teacher's moves because, in the beginning of the lesson, only two students were discussing the generalization. Kate made some significant moves to give other students access. I wanted Risa to state what she was seeing more clearly. I pointed out that in line 325, Kate asked the class if someone would restate Scott's idea and that she was met with silence. Up to that point, only Scott and Molly were talking. I asked, "What enabled other students to move into the ideas?"

Risa said, "When you ask a student to restate another student's idea."

Risa did not seem to understand which of Kate's moves were letting the rest of the class into Scott and Molly's thinking. However, I wanted to acknowledge what she said. "That is an important thing to do, to ask students to paraphrase another student's ideas, even though in this situation, the other students couldn't seem to understand the idea enough to say much."

Charlotte was looking somewhat agitated when she said, "Those kinds of moves are important, but it goes way beyond seeing the move as just asking the right questions. In the beginning of the case, you see the teacher has recognized

Finding Relationships in Addition and Subtraction

themes that come up in her classroom. She takes careful note of these themes and thinks about how to enlarge them. So she thinks about a theme and asks what is important about it for her students. Then she sets up situations in which they can state the idea, and she asks them what we all know that allows them to do this. The teacher is paying close attention to the math themes in the room that she wants to have discussed."

Charlotte was speaking about this passionately. I was not sure if I fully understood everything she was saying. I think she meant that if you just look at the superficial moves Kate is making—such as the kinds of questions she asks—you miss what makes the teaching effective. You need to pay attention to Kate's deep thinking that was guiding those moves.

I said, "I think Charlotte is saying something very important. Let's look at the first two pages of this case and notice what Kate is saying."

Charlotte said, "Absolutely. She is noticing the ideas that come up repeatedly in her classroom."

I suggested that the participants share what they noticed because I wanted to help other people see what Charlotte was seeing.

June said, "Kate notices patterns when the students work on 'number of the day.' It's like what we were doing before break. If you're subtracting and you add 1 to both numbers, the difference stays the same."

Yanique said, "They also noticed patterns in addition."

Madelyn pointed to the word problems. "She wrote those word problems so they could think about the generalization, 'If you add 1 to an addend, you add 1 to the total.'"

I said, "She took an idea that was emerging and created problems in order to nudge those ideas forward."

Phuong said, "Back around line 330, Molly and Scott were talking about numbers totaling to 100, and Kate then moved to numbers adding to 10. Moving 10 things around is easier than moving 100 things around. The smaller amount allowed them to generate more ideas than if they had used the larger number."

I responded, "Yes, there were two important things that happened there. First, Kate asked Scott to discuss the same idea using smaller numbers. She also suggested that he demonstrate with cubes. After that, many more students entered the discussion."

Denise interrupted, "I have got to say, I do love these cases, but where is the student who counts every dot on the cards and still needs to make the connection between the numeral 4 and the 4 dots that make up that quantity? What do we do about the students who are not there? All the cases in DMI only show students who are understanding wonderful things and are not showing struggling students. In these cases, we only see all this wonderful thinking—and it is wonderful—but that does not help us figure out how to help struggling students."

I do not know Denise well enough yet to understand what was behind all this energy. She is simply incorrect when she says that DMI cases never show struggling students. Assuming that she started out with Building a System of Tens, she need only go as far as Chapter 2 in that Casebook, and each chapter after that, to find struggling students. In this chapter of RAO, as well, some of the cases include struggling students.

However, that is not what I said aloud in the moment. Instead, I agreed that in any classroom, there are always students at every different level. It is just as important for a second-grade teacher to think about the next step for a child who is still "counting all" to add as it is to help students who are ready to articulate generalizations. "But one thing to consider is whether the struggling child gains anything from listening to discussions about generalizations. Imagine the students in the DVD clip we saw today."

Charlotte said, "I am thinking of Carlos in the DVD segment. He was the one who was counting to figure out 9 + 8. Maybe he still wants and needs to solve problems that way. However, he might still understand something when he listens to how Amalia and Manuel explain their strategies."

Madelyn added, "I would not want to leave it all at the level of abstraction if only two students can understand it. Going back to Kate, when it was completely verbal and when the numbers were large, only two students were talking. When Kate switched to smaller numbers and brought out the cubes, many more students were involved."

I said, "You know that not every student is involved in every discussion. In a lot of the cases, the teachers mention how many students are participating and make decisions about what to do next accordingly. What you do not want is to have the same students left out of every discussion. I would like you to think about Denise's question, and also these last points Charlotte and Madelyn have made. Think about what might be happening to students who are not ready to talk about the ideas but who are listening, and think about how Kate changed the discussion to let more students in."

Exit cards

Again, it felt like we had covered a lot of territory in those three hours. I wanted to find out where people were in terms of the work they had done with their own mathematics, as well as analyzing student thinking. So, I asked them to write about both in their exit cards.

All of the exit cards were positive. Many people wrote about how intrigued they were by Phuong's and James's representations for justifying the addition generalization. Several people were thinking about the two different ways to show subtraction on the number line.

Charlotte wrote about her concern that looking at teachers' moves is similar to "using key words to solve word problems" if you do not dig into the deeper intentions or agenda of the teacher. Vera asked how much math a principal needs to know in order to observe a math class. When I respond to their student-thinking assignment, I'll include some comments to Charlotte and Vera about these points.

Denise wrote, "In terms of the DVD and print cases, I think they are phenomenal, and I just love what the students do. I wish DMI would make DVDs and write cases about the students whose thinking is not quite there and ideas for moving these students forward. I know: Do the things you do with younger students. Use multiple representations to make ideas more accessible etc. However, I still end up being stuck." It seems that Denise has not been able to mine the DMI cases for her questions. Did she notice that Monica's discussion came from a simple subtraction problem that had stumped most of her students? Did she note what happened with Riley, a child who was usually completely lost during the class's math discussions? When I respond to her assignment, I will talk about this. I will also suggest that, in her next two student-thinking assignments, she write about struggling students so the two of us can think together about what she can do to help these students move forward.

Responding to the second homework

October 3

I looked over the generalizations—some in common language, some in algebraic notation—that participants had written from the cases. In general, the statements in common language were pretty good, and because participants already discussed them in whole and small groups, I did not respond to those. I was interested to see who was able to write the statements in algebraic notation and who gave it a try but had some of the conventions wrong. As I had done with Grace in the whole-group discussion, I responded to these participants explaining that I could see the logic in what they had written but showed them the correct convention.

As always, I learned a lot by reading these student-thinking assignments based on class explorations of odd and even numbers. One thing that struck me over and over again was the number of students who thought 5 is an even number. These students were in all grades from Kindergarten to sixth. In Denise's case, sixth graders were given the definition of odd and even—an even number has 2 as a factor, an odd number does not have 2 as a factor—and were told to model even and odd numbers by arranging square tiles in a 2 by x array. Denise described the work of one child who had created arrays for 2, 3, and 4 and began to work on 5. "He stopped and slid the fifth tile back and forth on the two rows of 2. Then he placed it in the middle of the two rows, drew the configuration in his book, and labeled it *5* and *even*. When asked about this, the student insisted, 'My 5 is even. I use it all the time. I do. I know it is even.'" So that is something for me to ponder. Why do many students come to the conclusion that 5 is an even number? Have their teachers been skipping over the definition of even numbers, assuming that students already understood them? Is there something so special about 5 that some students are drawn to think of it as even *in spite of* the definition? I will need to think about that.

I was also struck by the fact that some of the teachers in the group were still confused about the definition of *odd* and *even*. During our discussion in the first session, I thought we all agreed on the definitions of *odd* and *even*. However, it seems that the main confusion among these adults has to do with dividing a number "evenly." Because 9 things can be shared equally among 3 people, why is 9 not even? Although this issue was raised in our discussions several times, and I had also addressed this in the letter I wrote after the first class, it seemed as though there was still a bit of confusion. Perhaps for some participants, like Josephine, the issue is about learning in a second language. Josephine explicitly asked me to explain again what makes numbers odd or even. Yet, there were some others, such as Kaneesha and Vera, who became tangled up as they followed students' ideas when the basic issue was how even numbers were defined. Here, I share what I wrote to Josephine. I was trying to answer her question as directly as I could, while also appreciating what might be confusing to her.

Dear Josephine,

Even is a mathematical term that refers to those numbers that are divisible by 2 (without fractions or remainders). The relationship between even (that is, as a mathematical term) and 2 is simply that—it is how even is defined. In the English language, even has a lot of other meanings—"flat," "having no irregularities," "having the same height," "uniform," and so on. Those meanings are set aside when we are thinking about even numbers.

Just to make things more confusing, we sometimes use the term evenly in math class when we are not talking about even numbers. We can say, "3 divides 9 evenly," meaning that 9 divided by 3 does not have a remainder. When we talk about even numbers, we mean specifically those numbers that can be divided by 2 without a remainder.

There are different ways to think about even numbers: numbers you get to when you count by 2 starting at 0, numbers that can be broken into pairs, numbers that can be split into 2 equal groups without fractions or a remainder, multiples of 2, numbers with a factor of 2. If you take any whole number and think about all of these definitions, you will always come to the same conclusion—the number is either even or odd. These definitions are all equivalent.

You had also learned that the numbers 2, 4, 6, 8, 10 are even, and numbers with 0, 2, 4, 6, or 8 in the ones place are even. The definition goes back to dividing by 2. If you take any of those numbers that end in 0, 2, 4, 6, or 8, and you divide them by 2, there will be no remainder.

Your student, Jackie, was correct when she said that when you can count to a number by 2s, it is even. Tali corrected Jackie when she said that it is odd when you count by 5s. In fact, you can count to some even numbers by 5 (10, 20, 30, 40). It seems to be harder for these students to say what an odd number is, although they could tell you which numbers are odd.

Otherwise, students you interviewed are very clear about what even numbers are, and they are correct. As I have read the full set of episodes turned in by participants last week, I see that there are many students who are not so sure. I have read about quite a few students who believe 5 is an even number, and I have been thinking about why they come to that conclusion. Why we should make this classification of numbers based on 2 may not be obvious, and so some students think about other ways of classifying numbers. When that occurs, it is important to help them see that even numbers are about dividing by 2, but at the same time, acknowledge that they are noticing important characteristics of numbers, even if those characteristics do not have a special name.

Does my response help? Let me know.
Maxine

I will need to check with the teachers who have been confused about odds and evens to see if my explanation cleared up their confusion.

When I looked at Mishal's assignment, it seemed pretty incomprehensible to me—just sketchy notes. It appeared that she had written up a template before the lesson, which she tried to fill in with a few of the students' utterances while they completed the activity.

Mishal

Because I teach Kindergarten, I have decided to model my lesson on "Nadine, Case 2." Nadine's first-grade class was discussing odds and evens in October, and I will be working with my students in the beginning of the school year. It is because of these differences that I have decided to work with a small group of students who seem to have a good sense of numbers. We will be using manipulatives to work on the concepts of odd and even and to work through the discussion of adding numbers that are odd and even and seeing if there is a pattern.

The group will consist of Howie, Kathryn, Kent, and Selena.

We will count and write the numbers 1–20.
We will discuss what makes partners when we line up.

Let us pretend that a teacher has a class with one student. Will that person have a partner? (The teacher cannot be a partner.)

H: *There is only one.*

KA:

KE:

S: *No one would be beside him.*

We will try the number 2. Two would have partners.
What about 3? 680

H: *They would have 3 partners.*

KA: *No, you can only have 2 for partners.*

KE: *One would be left over.*

S: *Too many for partners.*

We began to make a picture chart to place the numbers. 685
We continued to do the other numbers 4, 5, 6, 7, 8, 9, 10 together.

Introduce the words odd and even.

Look at the chart that we made, any numbers that have all the numbers matched up are even . . . the numbers that have left over people will be odd.

Shade in the even numbers and see if the students can predict what will hap- 690
pen with the remaining numbers on the chart.

KE: *It looks like a pattern.*

Predictions:

H: KA: KE: S: *All agreed that it was a pattern, and they began to use the shaded areas to predict whether the numbers were odd or even.* 695

Allow the students to take the cubes (give each team two bags containing the cubes for 11, 12, 13, 14, 15, 16, 17, 18, 19, 20).

When the numbers were greater than 10, H and S began having a problem predicting or deciding whether a number was odd or even. Ke and Ka had no problem predicting and illustrating their numbers for the chart. 700

We did not have time to work on the following portion of the lesson. We will work on it during the upcoming week.

After the students have determined whether the number of cubes in their bags is odd or even, fill in the rest of the chart. Were their predictions correct?

What does it mean if we add things together? 705

H:

KA:

KE:

S:

Finding Relationships in Addition and Subtraction

Let us try some numbers, and see if we can make some predictions.

Let us take two numbers from the chart that are even numbers.

We take the numbers . . .

When we add them together, what kind of number is the answer?

What happens when we add two odd numbers?

What will happen if we add an odd number and an even number?

I think that the students had an easier time relating to numbers smaller than 10 because those are values/numbers that are more familiar to them at this time. Because it is early in the year, the concept of whether number values are odd or even is strange for most. I thought that it was interesting that the four students who participated performed at a more varied level than I had expected them to.

Fortunately, I had visited Mishal's small group just as she was explaining what had taken place in the lesson, and I had more to go on than what she had written. In my response to Mishal, I wanted to communicate my interest in her class and also to define for her more clearly what the expectations are for an assignment like this.

Dear Mishal,

Thank you very much for sharing this lesson with a small group of your Kindergartners. It is quite interesting to see what these young students can do with these ideas.

I do want to make sure that you do not feel pressured to push on ideas that are beyond where your little ones are. Most Kindergarten students have not yet begun to think about the operation of addition. However, if you have a sense that some of your students can engage with these questions and if you are curious about what they would do with them, by all means, check it out.

It was a very good idea to introduce to your students the idea of which numbers allow everyone to have a partner. As they thought about numbers in those familiar terms, you were in the position of giving them words for the two categories. Even numbers are those for which everyone has a partner and odd numbers are those for which one person does not have a partner.

It was also a good idea for you to structure the students' activity in a way that they could see a pattern. I believe you said that you had already been working with the class on patterns. This way, again, they could attach a familiar idea to new content.

Mishal, because I was present when you described the episode to your small group, I could make sense of your notes. However, if I had not heard you talk about the lesson, it would have been very difficult for me to interpret. You will be asked to do two more student-thinking assignments for this seminar. When you do this, I would appreciate it if you put more detail into your writing. For example, you gave your small group a clearer picture of what the activity was, and then you showed us how Kent and Kathryn kept track to be able to determine which numbers between 11 and 20 are odd and which are even. It is that kind of detail I would like you to include so that I can really picture what happened in the lesson when I read your case. I would also be interested in what Howie and Selena tried and where they got lost.

Some people take notes during a lesson to capture what students say and do; some people put a tape recorder on; and some people write notes as soon after the lesson as possible. Regardless of the strategy you try, when you actually write the episode, you need to replay the lesson in your mind and write down what happened so that a reader can understand.

Thinking back on what you told your small group and me, I am interested that the other students in your class also wanted to know about the ideas you were working on with Kent, Kathryn, Howie, and Selena. I wonder if it will make all of your students more curious about odd and even numbers throughout the year.

Thanks for sharing.
Maxine

Many of the participants' papers were quite powerful in terms of raising questions that their students engaged with and recording students' thoughts. In my responses to these, I chose to highlight some of the ideas the students presented, especially if students were using spatial representations such as cubes or tiles. I drew the participants' attention to how these representations served students' thinking. Many of the older classes defined even numbers as numbers with a factor of 2, so I discussed those ideas in my writings, as well.

In a few cases, I commented on pedagogical issues. It seemed to me that Risa fired questions at her students too quickly for thoughtful responses, and so I encouraged her to slow down. On the other hand, Lourdes' students were firing off ideas at a rapid pace, and I encouraged her to exert more control over the conversation. Phuong interviewed three students but left one behind, and so I tried to show her ways to work with that child.

Yanique's assignment is an example of a teacher who understands what it means to write a case, and at the beginning of the year, her students are beginning to learn what it means to pursue mathematical questions.

Yanique

Exploring Odds and Evens

The third-grade lesson I am going to write about focuses on what happens when odd and even numbers are added. After examining several sets of numbers, students are then asked to make generalizations about the behaviors of odd and even numbers.

750

755

760

765

770

775

780

785

To begin the lesson, I focused on making a chart of what ideas students 790
already had about odd and even numbers. They were asked to take what they
knew from previous experiences and expand on those ideas (what they learned
last year, solving doubles and halves problems). In the past, this exercise has
always been successful, but this group of students had a difficult time moving
beyond the idea that 0, 2, 4, 6, 8 are even because they "know it." 795

My question was posed as, "What can you tell us about odd and even num-
bers, and why do you think what you are saying is true?"

When I began charting students' ideas, we moved very slowly. I finally received
a response that even numbers are "equal." I chose to continue with this thought
because I needed to understand what students meant by that. 800

TEACHER:	*What does it mean that even numbers are equal?*
LIANA:	*Even numbers can be split and have the same amount.*
TEACHER:	*Can you tell a little more about that idea?*
LIANA:	*(She grabbed the cubes.) If I have 10 cubes like this (in a long stick), I can break it here and they are now equal. (She broke the stick into two groups of 5 and then lined them up next to each other.)*
ADAM:	*It also means that they can be put into 2 equal groups, it is fair. It does not just have to be 10; it can be something like 2, 4, 6, or 8.*
TEACHER:	*Adam, can you tell me why you listed those numbers?*
ADAM:	*I just know that those are even and they will be fair.*
TEACHER:	*Can anyone explain Adam's idea in a different way?*

At that point in the conversation, no one was ready to share. I am not sure
if it is because I did not ask more questions or they were just not ready for 815
that. As I paused, someone else let us know that 1, 3, 5, 7, and 9 were odd
numbers, but again, no one was able (or willing) to explain why. Although
students "knew" which numbers were odd and even, I was not convinced that
they truly understood why. I made the choice to move on to the combining of
these numbers in the hope that more students would discover ideas about the 820
behavior of these numbers. They had the choice to use cubes to work through
several addition problems, naming the sum odd or even and then writing
down what they noticed.

My question was posed as, "What kind of an answer do you get when you
add numbers? Pay attention to even + even, even + odd, and odd + odd." 825

At first, the students appeared to just use the cubes to find the sum to prob-
lems they were already familiar with. However, with a little encouragement,
they were then moving the cubes around and showing equal groups. Finally,

one student noticed that all the odd numbers had an "extra" cube. He was
referring to the cube that was left over or that didn't have a partner.

DANIEL: I was working on 4 + 3, and the answer is 7. I knew that,
but I used the cubes to check. When I put them together, I
had this leftover piece (pointing to the single cube on the
top of his stack).

TEACHER: So what does this tell you about the number 7?

DANIEL: I knew 7 is odd, but now I can see that it is the odd one. It
does not have another cube to be with. I was thinking of the
double problem when students were sitting with a partner
at a computer. (He is referring to a doubles/halves problem
done earlier in the unit about a class of 24 students in which
everyone can have a partner at a computer.)

Daniel went on to explain how he took 4 cubes and put them next to 3 cubes
and had the extra piece. This was an interesting strategy because many other
students tended to make one long stick and could not "see" that extra piece. I
continued to ask Daniel what this tells about adding an even and odd number
(something that he also mentioned in our conversation), and he stated that
he thought even + odd would always be odd. I left him to explore his conjecture more. After leaving Daniel, many other students who were listening in
on our conversation began to set up their cubes side by side to compare the
lengths.

Next, I sat down with Ana who seemed to be struggling with the concept
of the lesson. She wrote down her list of addition problems, solved them,
and then wrote if the sum was odd or even. She then took the cubes to check
the answers, and she got them all right and felt that she was done. My first
question to her was "What do you notice about the problems that you just
solved?"

ANA: I know all the answers because they were easy.

TEACHER: How did you decide if the answers to your problems were
odd or even?

ANA: I just know which numbers are even and odd.

TEACHER: Earlier today, someone in the class mentioned that even
numbers can be shared equally. I was also talking to someone about even numbers having partners. Can you talk to
me about these ideas?

ANA: Well, I think that even numbers like 2, 4, 6, 8 are even
because they are equal.

KHALED: (Another student in her group chimed in.) If you have 6,
you can make partners with each of them. Like 1 and 2, 3
and 4, and 5 and 6. (He was showing this with the cubes.)

| TEACHER: | *Does that mean that 6 is even?* | 870 |

| KHALED: | *Yes, because they were passed out evenly. If you had one more cube to make 7 then that would be left over and that would be odd.* | |

| TEACHER: | *So if 6 is even and 1 is odd and you add them together, what kind of an answer do you get?* | 875 |

| ANA: | *I think odd, because there will be that one left over (pointing to the cubes Khaled was using).* | |

| TEACHER: | *Good. What about if you add 6 + 2?* | |

| KHALED: | *That's 8 and even.* | |

| TEACHER: | *How do you know?* | 880 |

| KHALED: | *Because they each have partners (again showing with the cubes).* | |

| TEACHER: | *I would like you both to continue with these two ideas. So far you have worked with examples that have shown you even + odd = odd and even + even = even. Explore these ideas more and see if it works all the time.* | 885 |

By the end of the lesson, most of the students did begin to make some generalizations about even + even = even, even + odd = odd, and odd + odd = even. One student pointed out that she found more ways to get an even answer than an odd answer and had proved this using many examples. She had clearly labeled her paper and sorted the types of problems that made even or odd answers and then began to study those individually. I was impressed with her ability to work on this by herself until she came to a generalization. I also had two other students who were beginning to work on generalizations when subtracting numbers. They were finished early, so I paired them up to explore this idea.

I found it intriguing that so many students just accepted a fact to be true without being able to prove why (just listing the odd and even numbers). This inability to prove the fact is something we will concentrate on this year, and hopefully, students will begin to "self assess" and ask themselves questions about why numbers are working in certain ways. Overall, the class could tell if a number was even or odd, but had a very difficult time explaining without using the knowledge that 0, 2, 4, 6, 8 were even and 1, 3, 5, 7, 9 were odd because "they knew it." I did feel that by the end of this investigation the students were more comfortable manipulating the numbers and working toward an explanation by looking for commonalities in the behavior of numbers.

In my response to Yanique, I wanted to emphasize the thinking of her students who were working with images of odd and even numbers and were then able to engage with questions beyond answers to calculations.

Dear Yanique,

As I read the set of homework that was turned in to me last Wednesday, I found it very striking what happens when students begin to "see" what odd and even numbers look like and the power that students hold when they have that image to work with. Those who learned the list of even numbers and then were told to "just look at the ones place" didn't have a basis from which to reason. (Also, some students had incorrect ideas about what even numbers are.)

Daniel's "discovery" is an example of the power students begin to feel when they can work with these images of odd and even numbers. He says, "I know 7 is odd, but now I can see that it is the odd one." He is not just relying on having remembered 7 as belonging to the category called odd. He has an image of 7 that fits with the definition—and with that image he can connect the idea to previous work he had done on doubles and halves.

Daniel might still need to do some work to figure out how to reason about a whole class of numbers. For example, how would he model 3 + 6? Would he arrange his cubes in a row of 3 and a row of 6? He would need to do more rearranging in order to find that one cube that sticks out.

Khaled also has a way of thinking about even numbers that relies on an image of the definition. He sees the numbers grouped in pairs, and after talking about them that way, Ana begins to use that image, too.

I wonder why your students had such a difficult time moving into these images. It is as if they had learned these two categories of numbers as lists without connecting the categories to the attribute that defines them. Once they had those lists, there was not a reason to go beyond them. I found myself wondering whether they would have been able to identify larger numbers as odd or even.

Liana and Adam did have the idea that an even number is one that can be divided into two equal groups or an amount that can be shared in a fair way between two people. (All of this reasoning assumes whole numbers.) I wonder whether these two students used those images as they worked on the next part of the activity.

I think your instinct was correct to try to move your students into working with images of cubes to think about odds and evens. It was interesting that many of your students were building the numbers as single, long cube trains, which would not help them at all. However, once they saw the power of what Daniel was doing, they began to use his method.

Thanks for sharing this episode from your classroom.
Maxine

910

915

920

925

930

935

940

945

Detailed Agenda

Sharing student-thinking assignment

(35 minutes)

Groups of three

Organize participants into groups of three to discuss the reflections they wrote after examining the thinking of their students. While it is important that each participant have his or her paper read and discussed in the group, this is also an opportunity for each group to examine a set of papers for common themes and ideas. Announce to the groups that all of the papers should be read before beginning any discussion so that their conversation can reflect each person's ideas. They might begin by sharing what surprised them or what they discovered about their students' thinking. The discussion can also include similarities and differences among the ideas in the papers. Inform the participants that they will have 35 minutes for this activity, and they are responsible for making sure each paper is read and discussed during that time.

At the end of the activity, let the groups know you will be reading and responding to all of the papers and remind them of the procedure you established for collecting papers.

Discussion: Articulating Lola's students' generalizations

(20 minutes)

Whole group

As you call the group together for the next activity, let them know they will be continuing to examine student thinking with a focus on identifying stated or unstated generalizations throughout the seminar. While the work of the first session and the recent discussions were grounded in odd and even numbers, the group will now turn to a different math topic but will continue to focus on articulating and proving generalizations.

An important goal of this seminar is to support participants as they learn to identify and articulate the generalizations in the cases. As the participants read the cases to prepare for case discussions, they are also asked to write out the generalizations they see in students' work. Sometimes students are explicit about those generalizations, but more often they are implicit. This discussion, focused on Lola's case, provides the opportunity for participants to share what they have written and compare their statements with others in the group.

Begin by asking a few participants to share the generalizations they wrote based on Lola's case. At this time, elicit those statements written in common language. Statements might sound like, "If you have two cards and both of your numbers are larger than both of mine, you win," or "If we each have two cards and one of the numbers is the same, the other number determines which of us has the greater amount," or "If you add a number to a big number, the sum is more than if you add the same number to a smaller number." After each statement, give the group time to read and refine the statement, if necessary.

Ask what differences participants see among the statements. The statements might vary in terms of how general they are. For instance, are they expressed within the context of the card game or are the statements about numbers in general? The statements might also be grouped as two types: those that include statements about the operation of addition and those that do not.

Once similarities and differences of verbal statements have been discussed, ask if participants used symbols to express their ideas. One possibility might be "If a is greater than b, then $a + n$ is greater than $b + n$." If the ">" symbol is used, be sure to write both the symbol and the phrase *is greater than* so that all can become familiar with this notation. Invite participants to point out the connections between how the idea is expressed in common language and how it is expressed using symbols. (See the discussion of symbolic notation for Case 6 in "Maxine's Journal," p. 57.)

DVD: Subtracting 1 from an addend (20 minutes)

Whole group

Inform the group that they will be viewing a DVD of first graders exploring how the knowledge of one addition fact might help them solve another. Ask the participants to take notes while viewing the DVD so they will be able to discuss the thinking of each child. Display the poster you prepared with the names of these students from the DVD: Carlos, Amalia, Manuel, and Coleman. Tell the participants to pay close attention to the thinking of these four students.

After viewing the DVD, ask participants to explain what each child has done. Write these comments on the poster. Then allow the participants a few minutes to work with a partner to state the generalization that is implicit in the thinking of Amalia, Manuel, and Coleman. When the whole group comes together, solicit the generalizations first in verbal form and then in symbolic form. When symbolic statements are presented, suggest that participants substitute numbers for the variables so they can recognize how a specific numeric example is represented in the symbolic expression.

See "Expressing Generalizations" (pp. 52–53) for more information.

Math discussion: Proving generalizations behind our computational procedures (30 minutes)

Whole group

This math discussion provides the opportunity for the participants to explore computational strategies for solving addition and subtraction problems and the generalizations that underlie them. Also, this discussion introduces the use of the number line as a tool for expressing these mathematical ideas.

Begin by asking the group to do a mental math problem: 39 + 18. First solicit the answer, and then ask a few participants to explain their method of thinking. Next, explore a particular method in detail, that of changing 39 + 18 to 40 + 17. Present the equation 39 + 18 = 40 + 17 and invite participants to explain why it is true. Participants might say that 1 was taken from 18 and given to 39. Ask the group if they can apply a similar idea to a different problem, such as 47 + 24. (In this example, adding and subtracting 3 results in 50 + 21.) Ask, "Why does that work?"

Solicit a few responses by asking questions that require participants to state the generalization they are using. One statement might be, "If you reduce one addend by 1 (or 2 or 3 or . . .) and increase the other addend by the same amount, you do not change the total."

Pose the questions, "So, does it work for 39 − 18? Is that the same as 40 − 17? Can I take one from 18 and give it to 39?" Give participants a few minutes to work on their own or talk to a neighbor about the subtraction example. Once it is established that the principle is true for addition but not for subtraction, ask the group to make an argument to show that this statement is *always* true for addition. The arguments might include building cube structures, making story contexts, using a number line, or using symbols. See "Maxine's Journal" (pp. 59–61) for examples of such arguments.

The objective of this part of the discussion is to clarify that strategies and arguments are dependent on the operation being used. Solicit from the group how to modify this addition strategy to create a generalization that applies to subtraction. One possibility is illustrated by changing 39 − 18 to 41 − 20; that is, 39 − 18 = 41 − 20. For subtraction, if one of the numbers is increased, the other must be increased by the same amount to have the same difference.

Give participants a few minutes to talk to each other to explain this generalization. See "Maxine's Journal" (pp. 62–64) for more detail about using various contexts including number lines, to model subtraction situations.

Use the last few minutes of this discussion to pose the question from Case 9. "Let's say you start with 145 − 100. If you change 100 to 98, how does that change the result? That is, you know 145 − 100 = 45; how does that help you determine the solution to 145 − 98?"

If there is time, solicit a few comments from the group about this problem. If not, let the group know this question will be a part of their work in the case discussion.

Break
Break (15 minutes)

Case discussion: Students' generalizations and teachers' moves (55 minutes)

Small groups (30 minutes)

Whole group (25 minutes)

The cases in Chapter 2 present three main themes:

- What are the generalizations that underlie computational strategies for addition and subtraction?
- What are the forms of proof that students use to examine these generalizations? In particular, what are the roles of cube structures and diagrams in expressing generalizations and reasoning about them?
- What classroom dynamics and teachers' moves support students to reason about generalization?

Organize the participants into small groups, and distribute the Focus Questions for Chapter 2. Display the poster "Criteria for Representation-Based Proof" and remind participants to refer to these criteria as they examine student arguments. Let them know they will have 30 minutes for small-group discussion.

Focus Question 1 invites participants to share and compare the generalizations they wrote based on the cases in the chapter. Encourage group members to modify their statements as they discuss them. The participants will be handing in these statements for you to read, so inform them that it is important that their written statements represent their most current thinking. The other questions will build upon these generalizations.

Question 2 focuses on Monica's case (Case 9). Participants should first discuss the mathematics in the case and how ideas offered by one student are built on by others; then participants should consider the teacher's moves.

Question 3, based on Kate's case (Case 8), involves a similar kind of reflection. First, participants should examine and articulate the mathematics in the case, and then they should look at the teacher's moves to consider her intentions.

Question 4 addresses Maureen's case (Case 7). Participants examine how one student's use of cubes to express her ideas helps other students.

Questions 5 and 6 are based on Carl's case (Case 10). Participants examine the mathematical thinking of students, express the generalization underlying student thinking, and compare this with the mathematical ideas in Case 8.

Additionally, Question 6 raises the issue of forming a generalization with one set of numbers in mind and then having to decide if the generalization still applies when different kinds of numbers are to be considered.

Before the small-group discussion, remind participants that they might not have time to work on all of the questions. You might say, "If you are having a good discussion about the generalizations involved in these familiar computational strategies, what it means to you and to students to 'prove' such generalizations, and what kinds of teachers' moves support students working in this way, feel free to continue that discussion. It is understandable if you do not have the opportunity to address every question or case."

If the small-group discussions are going well, you might want to extend the time by 5 or 10 minutes. This will, of course, reduce the whole-group discussion time. On the other hand, if some of the small groups need a clearer focus, one strategy would be to call the groups together after 15 minutes and have a whole-group discussion on Monica's case using Question 2.

If you decide to discuss Question 2 at this point, ask participants to explain what Brian articulated and what Rebecca and Max added to the discussion. Solicit comments on Brian's explanation. "Is this a proof?" The objective of this question is to gather ideas about what a proof should be and how various representations model the operations. You can ask, "How does Brian's drawing represent subtraction?" or "How do other students make use of Brian's representation of the problem?" Once the group has discussed this question, they should return to their small groups to discuss the remaining questions in similar ways.

Focus the whole-group discussion on identifying the generalizations that underlie students' thinking about numbers and operations by discussing Question 2 (if you have not already done so) and then Questions 5 and 6. Ask questions such as "How does this argument fit the criteria?" and "How can you modify this argument to make it stronger?" If you have time, invite the participants to comment on the kinds of teachers' moves they noted in the cases that supported such thinking.

Homework and exit cards   (5 minutes)

Whole group

As the session ends, distribute index cards and pose these exit-card questions:
- We have had discussions based on the cases in the Casebook, your own cases, and DVD segments. What are you getting from these discussions?
- How are you doing with the mathematics in the seminar?

Distribute the "Third Homework" sheet and return the first homework assignment, including your responses.

Note: In order to make the RAO seminar productive for a wide range of participants, the agendas are designed to introduce symbolic notation gradually throughout the eight sessions, taking care that connections between symbols and the objects or actions they represent are carefully highlighted. However, some participants might be interested and able to expand their facility with symbolic representations at a faster pace than that set out in the agendas. If you have participants who are ready for such a challenge, then make available "Optional Problem Sheet 1." Let participants know this sheet does not replace the assignments that are scheduled. You can also inform them that they can request feedback from you at any time, but that these problems will not be discussed during the sessions. There are three "Optional Problem Sheets" included in the RAO materials. Each is intended to extend over two sessions. The other two sheets are included with agendas for Sessions 4 and 6.

Before the next session . . .

In preparation for the next session, read the participants' cases and write a response to each one. For more information, see the section in "Maxine's Journal" on responding to the second homework. Make copies of the papers and your responses for your files before returning the work. You should also read over the statements of generalizations written by the participants to get a better sense of how they express generalizations that they observe. Looking at the statements from each seminar session should help you track how your participants are advancing in their abilities to express generalizations in terms of words or notation.

DVD Summary

Session 2: Addition and Subtraction

Grade 1 with teacher Ana (3 minutes, 20 seconds)

On the DVD, there are some problems on the easel paper that are hard to read and are not a part of the taped conversation. It might be helpful for you to post the following examples, so participants will see what students were working on prior to the DVD taping.

$4 + 4 = 8$
$4 + 5 = 9$

$6 + 6 = 12$
$6 + 5 = 11$

We join the DVD as the teacher, Ana, writes $9 + 9$ on the board and asks for the answer. One student says the solution is 18 and another says it's 19.

Finding Relationships in Addition and Subtraction

Ana asks if the answer could be both 18 and 19 or if it must be only one number. Students say it has to be one number, but at the same time another student says the answer is 17. Ana summarizes by saying they have three possible answers: 17, 18, 19 and invites students to explain how they know which is correct.

Carlos says that he counts up from 9 and then 9 more to get a solution of 18. Amalia uses a derived fact: She knows 10 + 9 is 19, and so this answer is 19 minus 1, or 18. Manuel shares that it has to be 2 off 20, because 9 is 1 off 10 and you have to do that twice.

Then Ana writes 9 + 8 and asks if students can figure out the answer without working out the problem.

Coleman explains it has to be 1 less than 18 because 8 is 1 less than 9.

Posters for Session 2

1. Display the poster you made for Session 1: "Criteria for Representation-Based Proof"

2. Prepare a poster with the following four student names from the DVD. Leave room after each name. After viewing the DVD, record on the poster the participants' summaries of each student's ideas.

   CARLOS:

   AMALIA:

   MANUEL:

   COLEMAN:

Focus Questions: Chapter 2

1. Compare the generalizations you have written for all five cases. What is the same and what is different among your group's responses? During the discussion, make modifications to your statements so they express your current thinking.

2. Consider the passage from lines 450 to 475 of Monica's case (Case 9).

   a) What is the generalization that Monica's student, Brian, articulated? Did he prove it? If so, how? How did Rebecca and Max elaborate and refine Brian's idea?

   b) Once you have discussed the mathematics of this passage, turn your attention to the dynamics of this classroom. What do you note about the ways students and teacher interact?

3. In Case 8, the teacher, Kate, says at line 342 that her students "had learned something important about the operation of addition."

   a) Consider students' discussion from line 286 through line 340. Identify the flow of the mathematical ideas in this segment. What learning did students reveal?

   b) How did the mathematics students were working on in that passage, lines 286 to 340, show up in the computation later in the case?

   c) Consider Molly's comment at line 394. What question was Molly responding to? What was significant about what Molly said? How did Kate follow up?

4. What is the main idea that students in Maureen's class (Case 7) were working on? What did Sema demonstrate with cubes (line 101) and how did Connie and Ester connect with it?

5. Examine the class discussion in Carl's case from lines 603 to 628. Explain the thinking of these students. What images of the operation do they call upon in their work? How is the generalization the middle-school students are discussing the same or different from the generalization implied in the work of the second graders in Case 8?

6. What is the issue that Tammy brings up in line 635 of Carl's case? Does the generalization you wrote for this case still apply in the instances Tammy brings up? Explain why or why not.

Optional Problem Sheet 1

These problems will provide you with the opportunity to experiment with using symbolic notation. Feel free to turn in your work or questions on these problems at any time you would like comments or feedback. These problems will not be discussed during the regular seminar sessions. There will be two similar Optional Problem sheets available later in the seminar.

1. *In these expressions, both r and s are positive. If s increases, what will happen to the value of the expression; that is, will it increase, decrease, or remain the same? Explain how you know.

   a. $(r + s)$ b. $(s + r)$ c. $(r - s)$ d. $(s - r)$ e. $(r)(s)$ f. $(s)(r)$

   g. $[(r + s) - s]$ h. $\left(\frac{r}{s}\right)$ i. $\left(\frac{s}{r}\right)$ j. $(r + s - \left(\frac{r}{s}\right))$ k. $\left(\frac{r}{s}\right) - s$

2. Some expressions are not as consistent as those in Problem 1. Explain what happens to this expression as s increases from some very small positive number (e.g., 0.001) to some very large positive number (e.g., 1,000): $[r + s + \left(\frac{1}{s}\right)]$.

3. Consider each of the following statements. See if you can express in words what each one is saying. Then determine if it is true for all possible positive choices for a, b, and n. If the statement is always true, explain how you know. If it is not always true, explain why not. Use story contexts, diagrams or representations such as number lines to support your thinking.

   a. $(a + b) - n = (a - n) + b$

   b. $(a + b) - n = (a) + (b - n)$

   c. $(a - b) - n = (a - n) - b$

   d. $(a - b) - n = (a) - (b - n)$

   e. $(a - b) - n = (a) - (b + n)$

   f. $(a + b) = (a + n) + (b - n)$

   g. $(a + b) = (a - n) + (b + n)$

   h. $(a - b) = (a + n) - (b - n)$

   i. $(a - b) = (a - n) - (b - n)$

   j. $(a - b) = (a - n) - (b + n)$

---

*The expressions in problem 1 are adapted from materials by McCallum, Connally, and Hallett and used with their permission.

Third Homework

Reading assignment: Casebook Chapter 3

Read Chapter 3 in the Casebook, "Reordering Terms and Factors," including both the introductory text and Cases 11–14. As you read, note the generalizations that are present in the students' work.

Write out the generalizations in words and/or symbols and bring them to the next session. These will be shared with other participants and also turned in to the seminar facilitator.

Writing assignment: Questions about mathematics

What questions about mathematics have the first two sessions brought up for you? In about one or two pages, explain the mathematical questions you are wondering about as a result of this seminar.

REASONING ALGEBRAICALLY ABOUT OPERATIONS

Reordering Terms and Factors

Mathematical themes:

■ How do students demonstrate that switching the order of addends results in the same sum and that switching the order of factors results in the same product? (That is, how do they demonstrate the Commutative Property for addition and multiplication over whole numbers?)

■ What patterns occur when switching the order of the numbers in a subtraction or a division problem and how can we explain these patterns?

■ What similarities between the structure of addition and multiplication does this work suggest?

Session Agenda

Norms for learning	Whole group	15 minutes
DVD and case discussion: Reordering terms when adding and subtracting	Whole group Small groups	10 minutes 30 minutes
Math discussion: Reordering terms when adding and subtracting	Whole group	30 minutes
Break		15 minutes
Case discussion and math activity: Reordering factors, multiplying and dividing	Small groups Whole group	30 minutes 30 minutes
Preparing for the next student-thinking assignment	Small groups	15 minutes
Exit cards	Whole group	5 minutes

Background Preparation

Read

■ the Casebook, Chapter 3

■ "Maxine's Journal" for Session 3

■ the agenda for Session 3

■ the Casebook, Chapter 8: Sections 5 and 6

Work through

■ the Focus Questions/Math Activity for Session 3 (p. 115)

Preview

■ the DVD segment for Session 3

Post

■ Criteria for Representation-Based Proof

Materials

Duplicate

■ "Focus Questions/Math Activity" (p. 115)

■ "Planning for Student-Thinking Assignment" (p. 116)

■ "Fourth Homework" (p. 117)

Obtain

■ cubes

■ DVD player

■ graph paper

■ index cards

Illustrating the Commutative Property of Multiplication

Throughout the RAO seminar, participants will gain exposure to various techniques that can be used to represent an operation. Some of these representations are seen in the print and DVD cases; others, participants create for themselves. Representations can range from verbal descriptions of a situation to diagrams to arrangements of cubes. The action of each operation is incorporated into the representation in some way. Once representations are made explicit, participants can use them as tools for reasoning about the operations themselves. Some representations, however, may be better suited for some arguments than others. During this seminar participants will examine what each representation offers as a tool for reasoning about a particular operation and its properties.

In most communities in the United States, 3×5 is interpreted to mean 3 groups of 5. In other communities, 3×5 means 3 taken 5 times and is interpreted as $3 + 3 + 3 + 3 + 3$, or 5 groups of 3. In RAO, we will use 3×5 to mean 3 groups of 5 for the sake of consistency. Given this understanding, 3×5 might be modeled by

Skip-counting: 5, 10, 15

Jumps on a number line:

A verbal description of a situation, "I have 3 plates and each plate has 5 cookies on it."

A drawing illustrating groups:

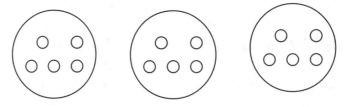

An array:

Each of these models illustrates the operation of multiplication. However, differences among these representations emerge when we examine how each of these models may be used to illustrate the Commutative Property of Multiplication, $a \times b = b \times a$.

For instance, the skip-counting or number-line models may be used to illustrate specific instances of the generalization that two factors may be exchanged and the product remains the same. For example, 3 jumps of 5 and 5 jumps of 3 both land on 15. However, these representations do not provide a general way to understand why it is that a jumps of size b will land at the same place as b jumps of size a. If you show a jumps of b, how can you be sure b jumps of a will land at the same place?

The array model offers an example of an argument that is not number specific. An array of a rows and b columns can be rotated to become an array of b rows and a columns. That is, the rotation transforms the array from a representation of $a \times b$ to a representation of $b \times a$ without changing the total number of cubes. This argument is not dependent on knowing the product of a and b.

The models that use a story situation or grouping diagram can also illustrate the general case. Consider this situation: There are 3 packages of cookies—one of chocolate chip, one of ginger snaps, and one of vanilla wafers—each package containing 5 cookies. This illustrates 3×5. However, if you distribute the cookies into 5 packages, each package containing one of each kind of cookie, the same cookies have been rearranged to illustrate 5×3.

This argument can be rephrased so it is not dependent on the number of groups or the number in each group. Consider a to represent the number of kinds of cookies and b to represent the number of cookies of each type. Then, grouping the cookies so each package has a single type of cookie results in a groups with b cookies in each group ($a \times b$). Grouping the cookies so each package has one of each kind results in b groups of a cookies ($b \times a$). Like the array, the story situation shows how you can transform a representation of one expression into a representation of the other.

For all of these arguments, the domain under consideration is positive whole numbers. When the domain expands to include zero, integers, or rational numbers, the reasoning about the Commutative Property will need to be revisited.

Maxine's Journal

October 8

Some years back, before I started to explore the Commutative Property with teachers and students, I would not have believed there was so much to think about. Since that door opened for me, I have found it to be fascinating. The questions that arise extend beyond the Commutative Property to include what happens when you change the order of the numbers with all four opera- | 5 tions or consider the result when you have more than two terms or factors. Last night was another example of the interest and engagement teachers experience when we pursue these fundamental ideas.

Norms for learning

Before we moved into the content of the session, I thought it would be useful to spend a few minutes discussing participants' experience with RAO thus far. In | 10 each seminar I teach, I find that it is important to pause around the third session to discuss the norms. I want to communicate to participants that the seminar is organized to promote their learning and that we should think about how to make it work for everyone. They need to understand that I will be responsive to them but also that they must take responsibility—both for themselves and | 15 the other participants. I was particularly concerned about helping some of the participants move beyond their negative associations with algebra.

I started out by saying, "We are now into our third session of the semi- nar. When we began, you all spent a few minutes talking about what you think algebra is. Some people reported that they had really liked the algebra | 20 class they took in high school; some people said it was a horrible experience. Whether you liked your past algebra classes or not, it is likely that the class did not include the idea that algebraic notation is a way of communicating ideas about the number system. In this seminar, we have been emphasizing ways of representing algebraic concepts other than with algebraic notation. | 25 However, as it comes up, we have also been looking at algebraic notation.

"Spend just a few minutes in small groups talking about how you are feel- ing about working with the algebraic concepts in this seminar. You may have talked about group norms in the DMI seminars you have taken in the past. In this seminar, too, we want to be concerned with how to be responsible both for | 30 individual learning and for the learning of the group. What should we keep in mind to be sure that this seminar will work well for everyone?"

After giving participants time to talk in their small groups, I brought every- one together for a whole-group discussion. Risa started us off by saying how much she learns from watching the DVDs. "We see the real-time interactions between students and teachers." ³⁵

M'Leah said, "I think it is really interesting to see the connections between the algebraic notation and the words that communicate the generalization. The algebra really means something." Yanique agreed.

I pointed out that at the beginning of the seminar, M'Leah and Yanique said they had liked algebra when they were in school. I wondered what was going on for those people who said that algebra was scary. I recalled that on the first day when we shared a single word associated with algebra, someone said, "Yuck." What is happening there? ⁴⁰

Charlotte said, "Well, I cannot say that I rejoice whenever I see algebraic notation, but I do start to have glimmers." ⁴⁵

June added, "I am OK if I can take the time to try it out with numbers."

I said that it is helpful to know how people are feeling about the notation, but the main point of the seminar has to do with the ideas—thinking about generalizations and ways to prove them, looking at the generalizations that arise in the elementary classroom, and seeing how students engage with them. "There is a lot of deep content for us to get to, and I want to make sure we are all able to support our own and one another's learning. What are the things we should be keeping in mind to make this seminar work well for everyone?" ⁵⁰

Several of the small groups had interpreted my initial question this way and were ready to offer some pointers. We ended up with the following list: ⁵⁵

- Complete homework.
- Cite evidence to support your claims about a case, and ask for this when others make claims. (Provide chapter, page, and line numbers.)
- Work to figure out what other people mean, and sometimes, paraphrase to see if you got it right. ⁶⁰
- Do not be afraid to speak. Be willing to share unfinished thoughts and also be willing to say when you do not understand.
- Make room for others to work things out their way and to share how they see things. ⁶⁵
- Use visuals—models and contexts—to illustrate the thinking being shared.
- Focus on understanding student thinking and not on our opinions of what the teacher is doing.
- Sit with new people each time to get to know everyone in the seminar. ⁷⁰

In fact, everyone in the group has turned in homework. I think that first point was to say how important that component of the seminar is—both for one's own learning and to be able to contribute productively to the group.

Similarly, participants sit with new people each session because I assign the groups. I believe Mishal named this as an aspect of the seminar that supports her learning. ⁷⁵

DVD and case discussion and math activity: Reordering terms when adding and subtracting

To move us into today's work, I began with a DVD clip of a second-grade class talking about what happens when the order of terms is changed in addition and then considering what happens when the order of terms is changed in subtraction. Although we were not going to take time to discuss the DVD segment as a whole group, I wanted participants to have these images while discussing the print cases. After the DVD clip, I said, "In small groups, you might want to discuss what you saw in the DVD. Mainly, I want you to have an image of young students talking about these ideas.

"For now, I want you to work on Focus Questions 1–5. The first four questions address the cases in which students talk about switching the terms when adding. Though in Alice's case (Case 13), as you recall, students were not all clear about which operations they were talking about. Focus Question 5 is about what does happen when you switch the terms of a subtraction problem. We will have a whole-group discussion about that before we go back to the case about multiplication."

Small groups

When I got to the table with Claudette, Antonia, and Phuong, they were looking at Kate's case. In this case, Kate had been working with her class on a problem involving a 20-cent coupon and a 10-cent coupon—did it matter in what order they were added? When the class answered with a resounding no, she asked them to prove that when you add two numbers, the order does not matter. This small group, looking at line 206, was struck by how Marissa had taken the idea of switching the order of addends in the context of their story problems but was now thinking about what happens when you break a number apart. Antonia said, "They are now thinking about switching order in a different context. They are also saying it does not matter when you add three numbers together, and they are just in second grade."

In contrast to the conviction held by Kate's second graders that order does not matter, "even if you do a billion numbers," some of Alice's third graders in Case 13 (Marina and Steve, at line 433) were not so sure. Phuong asked, "Why can't they generalize to all numbers? If something is so solid, once it is there with small numbers, why can't they generalize?"

It seems Phuong is still having a hard time moving into the perspective of the child. To Phuong, there is a single idea, and if you have it solidly, it is there for all contexts. She does not yet see that young students might maneuver quite confidently within the domain they generally inhabit. For these third graders, that comfort zone could be numbers into the hundreds, but the infinite world of numbers is still new to them. Furthermore, Phuong does not see it as a strength that Marina and Steve recognize the insufficiency of making a claim about an infinite class based on testing specific instances.

Antonia started to engage with Phuong about this idea. "You know, these third graders are different students from the second graders we read about. They might not have had such discussions before. Even so, no matter how solid they sound, when they are confronted with something new—like larger numbers or negative numbers—maybe they still need to go back and think again."

I pointed out that there is something very important about what Antonia said. Whenever we enter a new realm, we do need to think about our assumptions again. In fact, we are about to start looking at negative numbers, and so we will be needing to think about which of our assumptions based on counting numbers still hold, and which need to be revised. Next week we will be looking at a case in which students were checking to see if the "switcharounds" still work with negative numbers.

When I got to Leeann, Vera, and Jorge, I saw that Jorge had brought up an issue from his fifth-grade class about the Distributive Property. I listened a moment but then realized 1) they were not going to be able to get very far with his issue; 2) in just a few weeks we will get to a session that is set up to get underneath those questions Jorge is asking; and 3) if they keep going on Jorge's topic, they are not going to get to the ideas of this session. I mentioned my second and third concerns to the group. I said that, although Jorge's questions are important, I thought he would get more out of the seminar if he could focus on the work of this session. Otherwise, the group he was in would never have a chance to think about the content of this session and might possibly be lost for the next few sessions, which build on this one.

I was not sure what was motivating this group. Was Jorge trying to avoid something, or did he really believe he was on task by bringing these questions to his small group? Perhaps that does not matter. I moved away from the group so they could decide how to proceed. When I checked back with them a few minutes later, they were discussing Kate's case. Vera was asking, "Why did the teacher continue the discussion after students had already established, like around line 162, that order does not matter when adding?"

Yanique, Mishal, and Denise were looking at Case 13 in which Alice had asked her students to write about the rule she thought they had formulated together: *When you add two numbers together, you can change the order and still get the same total.* Mishal shared Alice's surprise, "Many students didn't realize they were talking only about addition." Denise pointed out, "But look, when they try it with subtraction, some of them see what is going on—that it is the same number, but a negative." Yanique added, "This is so cool."

When I got to Charlotte, M'Leah, and Kaneesha, they were into Question 5, figuring out what happens when you switch around a subtraction problem. M'Leah drew a number line and said, "Okay, $5 - 3 = 2$. I am at 5, I go back 3, and I land on 2. Now, $3 - 5$. I am at 3, and I need to go back 5. First, I will go back from 3 to 0. How much do I still need to go? I have to go 5 altogether, and I have done 3, and to figure out how many more I have to go, I do 5 − 3. So I go 2 beyond 0 and land on ⁻2. Is that always going to work?"

The three of them tried a few other examples to practice their argument and realized it would work for any two whole numbers.

Charlotte said, "If I know I will always end up the same distance from 0, if I subtract 5 − 125 or 125 − 5, well, it is easier to subtract 125 − 5. I know 5 − 125 = ⁻120. I could just think, 'I use up all the positive ones, and the rest is negative.'"

June, Grace, and Josephine had a different model, not using a number line. (At least to them, it felt like a different model.) Grace explained, "Let's look at 5 − 2. You have 5 things and you give away 2. You are left with 3. But if you have 2 things and you want to give away 5, then you still need to get 3 things."

They showed me their representation:

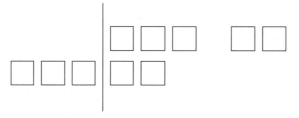

June said, "The top line is just 5 − 2. We drew that to show that giving away 2 things out of 5 leaves 3. The bottom line starts with 2 things. When 5 things are taken away, there are 2 on the right side of the line showing what you have; and 3 on the left side of the line showing what you still need in order to give away 5 things. In both versions, you see 5 and 2 and 3. However, in the bottom one, the 3 is in a different place. On the other side of the line, those 3 things are negative; they show what you still need to get."

Whole group

When I brought the whole group together, I asked, "Can anyone describe the regularity you are seeing when you switch the terms of a subtraction problem?"

Charlotte suggested, "They are both the same distance from zero."

Lourdes elaborated, "If the larger number comes first, the answer is positive. If the smaller number comes first, the answer is negative."

Leeann added, "The distance between the two answers is double what we got for an answer."

Leeann's observation is true. If the two answers are the same distance from but on either side of zero, the distance between them is twice that distance from zero. Although participants found that to be an interesting observation, my judgment was that it does not lead anywhere beyond that. So, I chose to clarify what Leeann said and then set it aside.

I said, "Let's check out these ideas with some specific numbers. 7 − 4 = 3. I think we are agreed that 4 − 7 = ⁻3." We explored the three statements by Charlotte, Lourdes, and Leeann against these numbers.

I continued, "Now, let's concentrate on Charlotte's and Lourdes's statements. When we switch the numbers around, we are still the same distance from zero, but one answer is positive and the other negative. Why does this happen? A lot of people are convinced this will happen no matter what numbers you use. How can you explain that?"

Jorge said, "It has to do with fact families." When I asked him to clarify, he said, "Well, 7, 4, and 3 are in the same family." However, he seemed unable to explain the negative result. *Fact families*, a term used in school mathematics, does not have to do with negative numbers but is about the relationship between addition and subtraction (which we will look at in Session 5). This was another idea that, in my judgment, would not help us with the content of this session, so we moved on.

Claudette said, "I used a number line," and came to the front to draw the following picture.

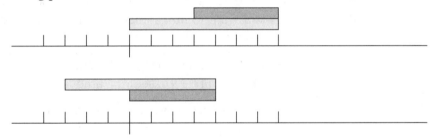

Claudette further explained, "Phuong cut some construction paper that made it clear to me. There is a strip that is 7 and a strip that is 4. If you are thinking about 7 − 4, you place one strip from 0 to 7. Then the other strip shows you go back 4, and you are left with 3. Then you can take those same strips and think about 4 − 7. You place the short strip to go from 0 to 4, and the other strip shows you go back 7. That same 3 is now on the other side of 0."

The group seemed very satisfied with Claudette's model. I said, "A lot of people are nodding. Claudette has shown us how this works with 7 − 4 and 4 − 7. However, we are saying this model works for all numbers. Can someone tell us how this model helps us think about all numbers?"

Denise volunteered. She explained that, no matter what numbers you use, you could have the same picture—just that the strips of paper might be longer or shorter. The top number line would show when the first number is larger. Then when the first number is smaller, you flip the strips over and push them down to line up the shorter strip with zero. That pushes the extra portion of the long strip to the other side of zero, but the amount is still the same.

I pointed to the Criteria for Representation-Based Proof poster and gave the group a few minutes to discuss how Claudette's representation, with Denise's extension, satisfies each of the three criteria.

After the participants' discussion, M'Leah said, "We were looking at number lines, too. The number lines are clear. However, if we apply the number line to a story problem, how does it work? I have 7 boxes and I need to give 4 away; I have 3 left. If you do the opposite: I have 4 boxes and need to give 7 away, so I will owe 3. How are these similar? We saw that in the first situation, we are left with the same amount that we owe in the second."

I asked if someone could see M'Leah's story problem on Claudette's number line. Risa said that if we are thinking about numbers to the left of 0 on the number line as owing boxes, then it is the same. Then I asked Grace to show us her model, with boxes, and M'Leah seemed to feel good about that.

Kaneesha observed, "I can see the 3 and ⁻3 in both representations, the number line and the boxes. They are really both the same."

I said, "Yes, when you can see two different representations coming together in your head, it strengthens the image. That is great."

Jorge said that his group thought about having and owing money. "If you have \$7 and want to buy something that costs \$4, then you will still have \$3. If you have \$4 and want to buy something that costs \$7, then you still need to get \$3."

I acknowledged this as another good example. "Looking at the number line, numbers to the left of 0 represent the amount you still need."

I was ready to move on. "Let's spend just a minute looking at algebraic representations. At the beginning of the session, we saw that for addition, we can say, $a + b = b + a$. So, how do we represent the generalization for subtraction?"

Lorraine offered, "$x - y$ = the difference."

I wrote $x - y = z$, and said, "Let's say z is the difference."

Lorraine continued, "So the other is, $y - x = {}^-z$."

I wrote that on the board and said, "When we use the same letter twice, it means that it stands for the same value." Then I wrote in "if" and "then" so that our statement read, "If $x - y = z$, then $y - x = {}^-z$."

Madelyn asked, "What tells us that x is the larger number?"

Risa called out, "Nothing."

I asked, "So what happens if x is not the larger number?"

James said, "That means z is a negative number. What is ${}^-z$? That is pretty confusing."

I acknowledged that it is pretty confusing. "Actually, if z is negative, then ${}^-z$ is positive. For example, ${}^-({}^-2) = 2$. Let's set that aside for now; we will get

98   Reordering Terms and Factors

more into that in later sessions. For now, I want to give you a vocabulary term: Because $z + {}^-z = 0$, we say that z and ${}^-z$ are *additive inverses* of each other."

Most people seemed to be familiar with that term.

I went on, "Also, I want you to look at another way to write out our idea." I put $(x - y) + (y - x) = 0$ on the board and asked everyone to write that down and to try it out with particular numbers.

I walked around to see what people were making of that equation. Lorraine said, "That is a pretty cool way of writing it. It doesn't matter which is larger, x or y. Either $x - y$ or $y - x$ is positive, and the other will be negative. They will always be additive inverses and when you add them, it will always come out to 0."

June told me she always needs to try out any algebraic notation with numbers, but now she is getting it. This is an important reminder to me: Whenever we write out a generalization in algebra, the participants should test it with particular numbers. That helps those teachers who are just learning to understand the notation make more sense of it.

I was just about ready to call a break when June raised a different question. "You said that when we use the same letter twice, it has the same value. Yet, I remember a couple of sessions ago you said that even when we use different letters, they can both have the same value. However, look at this." June was pointing to our statement, If $x - y = z$, then $y - x = {}^-z$. "What if $x = y$? Then it's not true anymore, is it?"

When I asked what the matter was, June continued, "Well if $x = 5$ and $y = 5$, then you have $5 - 5 = 0$ and $5 - 5 = {}^-0$. What kind of sense does that make?"

I still was not sure what the problem was, until more participants starting commenting on $5 - 5 = {}^-0$. Then, I understood. When I wrote on the board, $0 = {}^-0$, there were several comments of surprise. "How can zero be equal to its own opposite?"

I pointed out that mathematicians use the word *inverse* rather than *opposite* and added, "It might seem strange for 0 to be equal to its additive inverse, but it is. In fact, it has to be that way if we stick with the definitions." I wrote on the board:

$${}^-0 + 0 = 0$$
$${}^-0 + 0 = {}^-0$$

I explained, "We know the first statement is true because 0 and ${}^-0$ are additive inverses. We know the second statement is true because any addend plus 0 is equal to that addend. So in order to keep the system consistent, we have to say that ${}^-0 = 0$."

I told the group we would let that idea sink in over the next few sessions and called a break.

Case discussion and math activity: Reordering factors, multiplying and dividing

Once they came back from break, I asked participants to consider the different models the third graders in Case 14 used to illustrate "switch-arounds" for multiplication. Some students used skip-counting to show that two multiplication expressions result in the same number, while others created arrays. Two students, Sharon and Karen, had made 10 groups of 6, each group a different color. Then they redistributed the cubes, making 6 groups, each group with one cube of each of the 10 colors.

Small groups

As I visited the small groups, I heard Claudette express how intrigued she was by the model with the different colored cubes. "The different colors show so clearly the connection between the two multiplication statements, 10×6 and 6×10."

M'Leah, Charlotte, and Kaneesha were sorting out all three of the students' representations. They liked the skip-counting model because it showed the idea of multiplication so clearly. However, when they were asked to think about which models actually show the generalization about switching around the numbers, they looked at it differently. Charlotte said, "It shows you that you get the same answer for those numbers, but the representation does not hold the idea of switch-arounds. I can see it with the array more clearly."

Once the small groups finished discussing the case, I asked them to explore switch-arounds for division. When I got back to M'Leah, Charlotte, and Kaneesha a few minutes later, they had already seen that when you switch the numbers in a division problem, you get the reciprocal. Now they were thinking about how to demonstrate it with a story context.

M'Leah explained to me, "One story is, you have 12 candy bars to share among 3 students; for the other, you have 3 candy bars to share among 12 students." Then she showed me this diagram.

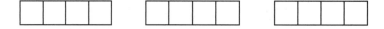

Her words were confusing to me, so I needed a moment to sort out what she was saying, but in the end I got it. You can look at that diagram in two ways. First, each square is a candy bar, and each group of 4 candy bars is what each child gets. That shows 12 candy bars distributed among 3 stu- dents, $12 \div 3 = 4$. If you look at that diagram for the second problem, then each long rectangle is a candy bar, and each square is one child's portion. That shows 3 candy bars distributed among 12 students, $3 \div 12 = \frac{1}{4}$.

Whole group

When I called the group together, I wanted to summarize the switch-around explorations for the four operations. Which operations give you the same answer when you switch the numbers around and which do not? Of course, everyone could name addition and multiplication as the operations that give you the same answer. So I introduced the word *commutative*. I acknowledged that some of them had heard the word before. Addition and multiplication are commutative; subtraction and division are not. Then I asked what people found when they switched the numbers in a division problem. All of the groups had found a pattern. Claudette came to the board and explained, "We looked at what you wrote for subtraction and wrote the division the same way. If $x \div y = z$, then $y \div x = \frac{1}{z}$."

Jorge said, "We wrote it as $\frac{a}{b} = c$ and $\frac{b}{a} = \frac{1}{c}$."

Both of these statements are appropriate ways of representing the idea.

James said, "Before we said z and ^-z are additive inverses. Is there a special name for c and $\frac{1}{c}$?"

I said yes and asked James if he wanted to guess what it was. He responded, "*Multiplicative inverse*?"

I acknowledged that he was correct and said another word for multiplicative inverse is *reciprocal*.

Then Risa asked, "What if you have the numbers 5 and 2? $5 \div 2 = 2.5$. Then this says $2 \div 5 = \frac{1}{2.5}$. Is that true? $2 \div 5$ is $\frac{2}{5}$."

I pointed out that we could do some arithmetic to see that, in fact, $\frac{1}{2.5} = \frac{2}{5}$. You get that by multiplying $\frac{1}{2.5}$ by $\frac{2}{2}$. Because $\frac{2}{2} = 1$, it does not change the value. Risa seemed satisfied with that.

Claudette said, "If you divide $x \div y$ and multiply it by $y \div x$, you get 1. It is the same as what you showed with subtraction, but now the equation is for division, and you multiply instead of add."

I wrote out $(x \div y) \times (y \div x) = 1$, and asked what people thought of that. Risa suggested you could also write it as $\frac{x}{y} \times \frac{y}{x} = 1$, and so I put that underneath the first equation.

I pointed out that when we are thinking about division we need to be very particular about the numbers we are talking about. Specifically, we cannot divide by 0, so neither x nor y can equal 0. Otherwise, these statements should be true for any values for x and y.

Then I asked the group to spend a couple of minutes examining those equations with numbers. Risa acknowledged that when an equation was written that way, you do not have to worry about whether the numerator is 1 or not.

Although I had been intrigued with what M'Leah had done with the story problems in small groups, I was not sure that enough of the participants would follow and felt we had gone far enough with the exploration of division. There was still more to do in today's session, so I turned to the cases.

I said, "I want to go back to multiplication and look at the representations the third graders used to justify their conclusion that switch-arounds for multiplication give the same answer. I wonder how you felt about each of those representations." 380

Before I elicited their comments, however, there was another point I wanted to emphasize. When I had taught this seminar before, some participants kept saying that they preferred the skip-counting representation because it made multiplication seem so clear. They did not seem to understand that the question was, Which of the representations supports the claim for *all numbers*? So I offered a few more words. 385

"When we talk about these representations, I want to emphasize that each representation helps us see some things but not all things. This is an idea we will be considering a lot in the upcoming sessions. So when I ask you how you feel about the representations, it is not just about which representation you like best for all purposes. I am asking, which representations allow us to feel assured that, when you switch the numbers in a multiplication problem, you will get the same answer—not just for the numbers we check, but for all whole numbers?" 390

395

June started us off by saying, "I like the skip-counting model best."

When I asked why, she said, "It is clearest in the way it shows multiplication. I think the array model is harder to see."

This is exactly what I had run into before. June likes skip-counting on the number line because it fits her image of multiplication—but she is not necessarily considering how one instance can be extended to support the claim for all numbers. 400

I found the chart that had Claudette's representation of the number line to show that $4 - 7 = {}^-(7 - 4)$ and reminded the group of Denise's description of how the strips could be stretched or shrunk to accommodate any pair of numbers. "Is there a way to do the same thing with the skip-counting model?" 405

I drew a picture of a skip-counting model for 4×7 and 7×4 and asked if anyone could show us how this model could be extended to show it works for all numbers.

Denise pointed out that we could stretch the whole thing. She suggested, "Let's stretch it to twice as long to see what happens." 410

After she showed us what she was thinking, she said that it does not help. Stretching out the number line just stretches out each of the jumps. It seems to show that $7 \times 8 = 4 \times 14$, and that will not help us.

Reordering Terms and Factors

Lorraine then said that she found the array model most convincing. "You do not have to change the array at all, just its position."

I asked how that was different from skip-counting, and Lorraine explained, "For skip-counting, it seems that you can only show that for the numbers you pick, if you reverse the numbers you get the same product. It does not show that it will work for all numbers, but only the numbers you happen to pick. However, with the array, you can see it will work for all numbers. If you make an array for any multiplication problem, you do not have to change the array at all. Just turn it, and it will show you the other problem. The answer has to stay the same."

Yanique said, "You do not even have to turn the array. It just depends on how you look at it."

We looked at an array together, and Yanique explained. "It just depends on your perspective. You can say that each row is a group and the number of rows is the number of groups. Here, you have 4 × 7, or you can say that each column is a group and the number of columns is the number of groups. So you have 7 × 4. You can see that it has to work for any multiplication problem."

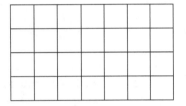

M'Leah said, "With the skip-counting you could not visually see the connections. It is just that the numbers you use end up at the same place. I can even imagine feeling suspense as you test it out. Will the numbers really land at the same place? With the array, you *have* to have the same number of squares. That cannot change whether you think of a row as a group or if you think of a column as a group."

Charlotte added, "The array can be any length and width. Nothing about the quantity changes when you turn it."

I let the discussion end here.

Preparing for the next student-thinking assignment

Before the end of the session, I put participants into approximate grade-level groups to think about their next student-thinking assignment. The handout I distributed was intended to communicate the kinds of questions they should be asking themselves before they actually perform the work with their students. I wanted them to be clear about the task that they would pose, why they chose that task (that is, which mathematical idea they expected the task to raise for their students), and key questions they might ask their students as they work.

When I set the group to work, the energy was high. They were very interested in finding out what their students might do with some of the questions we have been exploring in the seminar. Even though it was 7:15, the end of the session, many participants stayed to extend their discussion. I interrupted them to ask that they write their exit cards, and then they continued. A few were still there when I left the room.

Exit cards

For the previous homework assignment, participants had written about the mathematical questions the first two sessions had brought up for them. For their exit cards, I asked them to write an addendum: Were any of these questions resolved, and did new questions arise for them?

As I look over their exit cards, it seems as if this third session landed people in very different places. Charlotte wrote, "The insights of today totally shifted the entire course for me." She went on to write about how much more confident she feels now with the understanding that we are working at developing models to support how we see generalizations. The norms discussion at the beginning of the session, she said, helped, too.

Several people chose not to write about mathematics but instead expressed their curiosity about how their students will think about such issues as commutativity. Claudette wrote about how the idea of commutativity of addition recently arose in her first-grade class.

June wrote, "I do not quite care for the array model because 4×5 looks too much like 5×4, and that can be confusing. Skip-counting makes the differences clearer." It seems she does not yet understand how we are trying to use these representations to justify general claims.

I am concerned about Josephine who wrote, "I did not go too far today. I just do not get it for now. I know I will. I will work at it."

Responding to the third homework

October 13

Before I write about the homework, I do want to describe an interaction I had with Josephine during the week. It was in the context of a different professional development setting that Josephine happened to attend. In this setting, participants were exploring the different kinds of situations that are represented by subtraction (e.g., comparisons, finding a missing change, and so on). One problem was, "A baker had 16 sticks of butter in the beginning of the day and 11 sticks of butter at the end of the day. How many sticks of butter did she use?" One participant's solution involved a number line.

```
0                                    11              16
|__|__|__|__|__|__|__|__|__|__|__|__|__|__|__|__|
```

I do not need to go into the discussion of the problem, except to say that at one point Josephine raised her hand and explained, "You can see the stick of butter as the space before the number. The space between 0 and 1 is the first stick of butter. The space between 15 and 16 is the sixteenth stick of butter." | 485

Later she came to me and said, "I am so excited. I now understand what was so confusing to me." I now realize that Josephine had never used a number line before, and with all of our discussions in the seminar that relied on that representation, she had been completely lost. Her revelation was being able see the interval from one number to the next as a countable unit. It seems that Josephine has a very weak mathematical background, but she has a strong sense of herself as a learner and can identify quite clearly what she has been confused about and has come to understand. | 490 495

Back to the homework: Participants had been asked to write about the mathematical questions that had arisen for them in the first two sessions. Lourdes's questions began with the cases of students' thinking, but her questions led to serious mathematical issues. | 500

Lourdes

Questions About Mathematics

I am wondering about proving rules. In one case, some students insisted that there were numbers "out there" they had not tested and were, therefore, unwilling to accept the rule. Other students were satisfied after a few tests. In my fifth-grade class, many students are satisfied with a rule after just two tests. I imagine that mathematicians can create computer programs consisting of innumerable tests to justify new ideas or theories. | 505

Sometimes students come up with the rule: If you multiply a number, it will always get bigger. For the whole numbers 0–12 there are exceptions of 1 and 0. So I am wondering if math rules can have exceptions the way the rules of English pronunciation can. | 510

I am also curious about whether the number of tests is a developmental factor. When do students decide they do not have to test every single number? It seems to be related to the amount of experience they have with patterns or systems. | 515

Related to these rules is the zero. No exceptions can be tolerated for zero. I like testing assumptions such as "zero is odd." $0 + 3 = 3$, which is odd. This equation cannot be true because two odd numbers added together are supposed to equal an even number. Therefore, the assumption is wrong. So, by agreement, 0 is even because when 0 is even, all the rules work. Why does the | 520 *system have to be consistent? Languages are not. What would happen if we had to memorize some inconsistencies?*

My third concern is the way people internalize math. I have many students who just do not know facts such as 8 + 9, or 6 × 7, because they do not have available strategies in their minds. They do not count on their fingers or break the numbers apart. They do not make connections. A strategy suggested one day is forgotten the next. The very people who would benefit from some varied strategies cannot line up any connections. What kind of experiences in math could help work around this problem? Many teachers are concrete and sequential in their approach to math. This approach seems appropriate, but obviously it is not enough. Getting these students to talk, to articulate their thinking, might help.

Why does "invert and multiply" work for division of fractions? Is there a model that can be set up?

At what age can students understand generalizations represented by letters? Are we able to teach this?

Lourdes has asked important questions about generalization and justification. However, from what she wrote in her first three paragraphs, it seems that she has not yet taken hold of what a mathematical proof is and how it offers more than making a claim based on several—or many—examples. This lack of understanding was the main point I chose to address in my response to her.

In the fourth paragraph, Lourdes is talking about mathematical reasoning. If even and odd numbers are defined in terms of counting numbers—e.g., a number is even if it can be broken into pairs—can you say whether 0 is odd or even? She shows that if you start with the assumption that 0 is odd, that leads to inconsistencies, and therefore, 0 must be even. This reasoning leads her to another question about the commitment to consistency in mathematics. This is another important point to address.

Dear Lourdes,

You have lots of really good questions. I will not be able to address all of them in this response, but you should hold onto those questions and keep thinking about them.

One of the things we are looking at in this seminar has to do with ways to prove that something is true, that it will always work, and when testing it lots of times is insufficient. Take, for example, some of those arguments that were made about odds and evens on the first day of class. If we are working with a model of even numbers being those numbers that can be broken into pairs and if you add two even numbers, you have one set broken into pairs and another broken into pairs. When you join those two sets, you have a new set broken into pairs—another even number. At that point, we do not have to test lots of numbers. We know that any two even numbers added together will produce another even number. Similarly, when we work with an image of odd numbers being a bunch of pairs with one left over, when you add two odd numbers, the leftovers come together to make a pair, so again, the total is a big bunch of pairs. That tells us that the sum of two odd numbers will always be even.

In the last class, we were looking at students' models of multiplication, considering whether you can always change the order of the factors and maintain the same product. The question was, Which of those models allows us to picture in our minds what happens when you switch the order of any two factors? In that discussion, I realized that not everybody was so comfortable with the array model of multiplication—although that is a model that allows you to look at the multiplication in both directions, a × b and b × a.

Frequently, when we work on mathematics, we start to notice regularity, seeing how something works in several cases. We might test it a few more times to gain confidence in our observation. However, it takes more than checking several times to show that it will always work. That is when this other kind of proof is needed.

In general, testing things lots of times satisfies young students. In my experience, by about third grade, some students become aware of the infinite nature of numbers and start to realize that testing something a lot of times does not guarantee that it will always work. Some of these students might believe that one cannot ever say something is always true. Eventually, they should come to see that there are these other kinds of proofs (like what I have described above about odds and evens) that allow us to make claims about always. I am not saying that all third graders get there—but that seems to be the age when some students start to think in these terms.

You asked about students who do not have ways to think about varied approaches and also cannot remember their multiplication facts. You suggested that maybe getting these students to talk would be a way to help. I say, absolutely, try to get these students to talk about the way they are viewing the mathematics. It might be the case that they do not have any way of thinking about what multiplication is. If that is true, then you need to help them picture groups of things and how those groups are related to multiplication. Do they have images for addition and subtraction? Sometimes you have to go pretty far back to discover where these students feel solid. Then you need to help them work forward from there. It is not easy, but it seems to me that that is how to get started.

About the "invert and multiply" rule: I think that question is related to what I wrote in response to your first question above. You might look back at some of the dividing fractions problems from MMO—like Wanda's cake problem. Draw a diagram of the problem to show $4 \div \frac{3}{5} = 6\frac{2}{3}$. Then see if there is another way to look at that same diagram to show $4 \times \frac{5}{3} = 6\frac{2}{3}$. If you can see both expressions in the same picture, that would be a way to show "invert and multiply" for that particular case. The next challenge would be *to see if you can use that diagram to visualize what happens with other numbers. Can you go from there to consider all fractions (larger than 0)?*

There is a way to prove it relying on the consistency of the number system or just by using symbols. You can think about $4 \div \frac{3}{5}$ as "What do you multiply $\frac{3}{5}$ by to get 4?" You might play around with that, and if you are still curious about it after Session 5, bring that question to me again.

Maxine

P.S. Following your question, Why does the mathematical system have to be consistent? I talked to some of my mathematician and mathematics-educator friends. I think the answer to your question is that is what mathematics is! It is about being able to discover new things by following a line of reasoning. Although we begin to learn about mathematics through our interactions with the world—counting objects, combining, separating, grouping, sharing, comparing, and so on—as one continues in mathematics, it gets more and more distant from anything we can touch. At that point, logical consistency is what is used to make sense of things. Still, you have asked a great question that is worth continuing to ponder. You might keep your question in mind as you continue to work on the mathematics of this seminar.

610

615

620

Reordering Terms and Factors

Detailed Agenda

Norms for learning

Whole group

Now that the group has been together for two sessions and participants have had a chance to work together, it is a good time to collect ideas from the group about how to collaborate to assure that everyone benefits from the discussions. While you should remind the group of the norms that are expected during all DMI seminars, there are also some special issues that the ideas of this seminar might bring up.

For instance, some in the seminar might have had positive experiences with algebra when they were students, whereas some did not; some participants might be comfortable expressing ideas with symbols, whereas others are just now learning to use symbols to express their ideas. In either case, it is likely that the ideas of this seminar are causing all participants to wonder about their previous experiences with algebra and the connections between algebra and the arithmetic they teach every day. You may want to highlight a few comments from the preseminar writing or exit cards to initiate this discussion.

This is an opportunity to think about how the seminar has been working and to compile a list of seminar guidelines to help make it a positive learning experience for all. Examples of statements might include

- Allow time in small-group work for individuals to think before talking.
- Be prepared for sessions—everyone should have something to contribute.
- Listen carefully to take in another person's ideas.
- Find ways to disagree without being disagreeable.
- Start and end each session on time.
- Be open to a new idea or perspective.

Set aside just 15 minutes for this discussion. The list does not need to be comprehensive or complete. The activity is designed to help participants recognize that reflecting on their own learning process and considering the dynamics of the group is a part of the seminar. Some facilitators post the list in the meeting room and encourage participants to add to it as the seminar continues.

DVD and case discussion: Reordering terms when adding and subtracting

Whole group (10 minutes)

Small groups (30 minutes)

The DVD clip for this session provides images of students engaged in discussions similar to those in the print cases. The narrated 8-minute DVD segment shows second-grade students working to explain how they know that $23 + 2$ is the same as $2 + 23$. Near the end of the DVD clip, the teacher, Karen, asks them to consider $7 - 3$ and $3 - 7$. After the clip, suggest that participants keep the images from the DVD in mind as they turn to a discussion of the print cases.

These cases explore how students notice and develop ideas related to commutativity of addition and the non-commutativity of subtraction. Also, the cases offer examples of teachers' moves that encourage and support students as they examine and question their own ideas and begin to develop a sense of what it means to make an argument for a claim about an infinite set. Distribute the "Focus Questions/Math Activity" Sheet to participants. They should work on the first five questions at this time. (Questions 6 and 7 will be the basis of discussion later in this session.) When there are 10 minutes left, ask the small groups to move to Question 5, if they have not already done so.

Question 1 examines the teacher's questioning of Kindergarten students in Case 11 to help them explore their generalizations and consider the situations in which the generalizations apply. Some students in the case are able to reason abstractly about the numbers; others are still seeing the situation in the context of the checkers.

Question 2 continues the examination of the teacher's moves in Case 12 that support and encourage students to think more deeply about their generalizations and to determine how they could prove the generalizations to others.

Question 3 provides the opportunity for participants to discuss the ways students in Case 12 are thinking about addition and commutativity, in particular as students extend the generalization to more than two addends.

Question 4 addresses the variety of ways students in Case 13 think about what happens when the order of the numbers is switched in a subtraction problem.

Question 5 provides the opportunity for participants to explore for themselves what happens when the order of the numbers in a subtraction problem is switched. During this time, circulate around the room and encourage the groups to use a variety of representations (number lines, cubes, diagrams) to express their thinking. As you interact with the small groups, note what representations they are using and consider how to organize the whole-group discussion.

Reordering Terms and Factors

Math discussion: Reordering terms when adding and subtracting (30 minutes)

Whole group

The whole-group discussion focuses on Question 5: What are the patterns the participants noted when the numbers in a subtraction problem are switched and how can they explain those patterns? Some participants may be surprised to realize that even though subtraction is not commutative, the answers to the "switched-around" problems are related. Begin by posing a question the group can explore together: "What can you show about the relationship between $5 - 3$ and $3 - 5$?" "Why is it that one is equal to 2 and the other equals $^-2$?" Remind the group that, although they are using this example to express their reasoning, they are working on an idea that applies to all pairs of numbers; that is, why is it that if $a - b = c$, then $b - a = {}^-c$?

Request that some of the participants express their ideas. Begin with explanations that use verbal descriptions, diagrams, cubes, and number lines. As these are shared, provide time and ask questions so that participants examine each argument and explore the connections among them. See "Maxine's Journal" for Session 3 (pp. 94–99) for examples.

Introduce the term *additive inverse* to name the relationship between a and ^-a. Use examples (3 and $^-3$, 4 and $^-4$) to illustrate that two numbers are additive inverses of each other if their sum is 0.

Conclude this discussion by expressing the rule with symbols. Use this form: $(a - b) + (b - a) = 0$ rather than $(a - b) = {}^-(b - a)$. Writing it this way accomplishes two goals:

1) It avoids $^-(^-2)$, an issue that can arise with the second notation, for instance, if $a = 4$ and $b = 2$. (See the Facilitator Note below for more information.) Session 5 is focused on making meaning of negative numbers, and it is best to set aside this issue until then.
2) An analogous rule for division is easier for participants to formulate if they understand this notation.

Thinking about $^-(^-2)$

In Session 4, RAO participants work with models of integers as they begin to examine the meaning of negative numbers and to explore addition of integers. In Session 5, they extend that work to subtraction of integers.

In this session, negative numbers arise from reversing the terms of a subtraction problem, and the term *additive inverse* is introduced. Addition and subtraction are binary operations, meaning they operate on two numbers. Finding an inverse is an unary operation; that is, one can think of finding the additive inverse as an operation on a single number.

The additive inverse of x is ^-x which, when added to x, gives 0. When $x = 3$, the additive inverse is $^-3$: $3 + ^-3 = 0$. And when $x = 7$, the additive inverse is $^-7$: $7 + ^-7 = 0$.

Because $(0) + (0) = 0$, 0 is its own inverse.

When $x = ^-2$, the additive inverse, $^-(^-2)$, is the number that, when added to $^-2$, gives 0. That is, $^-2 + ^-(^-2) = 0$. However, we also know that $^-2 + 2 = 0$. Thus, $^-(^-2)$ must equal 2.

Break   (15 minutes)

Case discussion and Math activity: Reordering factors, multiplying and dividing   (60 minutes)

Small groups (30 minutes)

Whole group (30 minutes)

Participants should return to the same small groups. Inform them that they will be doing some similar thinking but will now consider the operations of multiplication and division rather than addition and subtraction. Participants will use Focus Questions 6 and 7 as the basis for this work.

As you circulate among the small groups, check that participants are correctly interpreting Question 6. This question, based on students' work in Case 14, invites participants to analyze the various arguments students offer to illustrate the Commutative Property of Multiplication. See the Facilitator Note, "Illustrating the Commutative Property of Multiplication" (p. 90) for more information.

Question 7 is an opportunity for participants to examine what happens when numbers are switched around in a division problem. The goal of this discussion is to help participants recognize the similarities and differences between the earlier work they did examining subtraction and the new work of examining division. You might want to suggest that groups work with number pairs that have no remainder (such as 6 and 2 or 12 and 3) so that complications in the calculations do not hide the basic relationship. Also, remind the group to consider diagrams and story situations to support their thinking. See "Maxine's Journal" for Session 3, (pp. 100–103) for examples.

Begin the whole-group discussion by inviting participants to act out with cubes some of the student arguments for the commutativity of multiplication. Ask how the cube models apply to all pairs of whole numbers and not just the specific numbers in the examples. You might also solicit comments from participants on whether students in the cases were making general arguments or illustrating specific cases. Questions such as "What is it that students say or do in the cases that provide evidence for your conclusions?" or "What do you look for in student work to help you decide?" or

"What questions might you pose to students as you examine this?" might be appropriate.

Then initiate a discussion about the operation of division. "What did you notice as you considered pairs like 12 ÷ 3 and 3 ÷ 12?" Focus the conversation on the analogies between addition and multiplication; for instance, a statement such as $(a ÷ b) × (b ÷ a) = 1$ can be compared to the earlier statement $(a - b) + (b - a) = 0$. [*Note:* $(a ÷ b) × (b ÷ a) = 1$ can also be written as $\frac{a}{b} × \frac{b}{a} = 1$. That is, $a ÷ b = \frac{a}{b}$ and $b ÷ a = \frac{b}{a}$.]

Define the term *multiplicative inverse* at this point to name the relationship between $\frac{a}{b}$ and $\frac{b}{a}$. Use examples such as $\frac{2}{3}$ and $\frac{3}{2}$ or 4 and $\frac{1}{4}$ to illustrate that a pair of numbers are multiplicative inverses if they multiply to 1.

Preparing for the next student-thinking assignment   (15 minutes)

Small groups

Distribute the "Fourth Homework" Sheet and the "Planning for Student-Thinking Assignment" Sheet. You might want to place participants in grade-level groups for this activity. In small groups, participants will design an activity or a set of math questions to pose to their students. Invite participants to think strategically about how to motivate a discussion that will move students toward thinking about generalizations. Suggest they use the planning sheet as a way to strategize about the questions they will ask their students. Let them know they will meet with the same small group in the next session to share their cases.

Note: Record the names of the participants in each group so you can use the same groupings for the next session.

Exit cards   (5 minutes)

Whole group

As the session ends, distribute index cards and pose these exit-card questions:
- What math ideas addressed in this session are clear to you and what topics are you still wondering about?
- What comments do you want to make about the norms for learning discussion?

Before the next session ...

In preparation for the next session, read participants' papers, generalizations, and exit cards and write a response to each participant. See the section in "Maxine's Journal" on responding to the third homework for more information. Make copies of both the papers and your response for your files before returning the work.

You should also read over the statements of generalizations written by the participants to get a better sense of how they express generalizations that they observe. Looking at these statements throughout the seminar sessions should help you track how your participants are advancing in their abilities to express generalizations in terms of words or notation.

DVD Summary

Session 3: Commutative Property

Second-grade class with teacher Karen (8 minutes)

In this narrated DVD segment, second graders are discussing the result of reversing the numbers in addition or subtraction problems. First, the teacher solicits examples of two addends that total 25. After gathering some examples, the teacher focuses the conversation on the question, "Is 23 + 2 the same as 2 + 23?" Student arguments include comments such as "You didn't change the numbers" or "You didn't add any or take any away." They also include comments about switching the positions of two stacks of cubes.

The teacher points at a problem on the poster 176 + 266 = 442 and asks if the answer to 266 + 176 is also 442 and why. One student explains that because the numbers are the same, the answer will be the same. Students offer, "You are just switching them around." and "Everything stays the same."

Then the teacher asks students to consider 7 − 3 and 3 − 7. Some students say it cannot be done, offering arguments such as "7 is not in 3" or that 0 will be repeated when you count down as in 3, 2, 1, 0, 0, 0, 0. One boy says the answer will be ⁻4.

Focus Questions/Math Activity

1. In Case 11, line 54, Lola asks her class, "Is this getting the same total no matter which order you counted something special about the checkers, or would it work with anything?" What was Lola alluding to with this question? What did Amber reveal in her response?

2. In Case 12, line 129, Kate says to her students, "When you are doing your explaining, I want you to pretend that Mr. Valen is standing here and that he doesn't believe you so you've got to be really careful to convince him. Don't assume anything. Say all the things you need to prove it." Why did Kate mention this again? How does Kathleen respond?

3. In Case 12, line 194, Kate writes, "I knew that Corey, Marissa, and Scott had thought about this in a more generalized way so I called on them to continue the discussion." Examine the thinking of these children as they generalize the addition rule to include three or four more addends.

4. In Case 13, lines 292–293, Alice defines the class's switch-around rule as "When you add two numbers together, you can change the order and still get the same total." Although the phrase specifically mentioned addition, some students still thought about subtraction. What did students say about subtraction?

5. What happens when you switch around the numbers in a subtraction problem? What are the regularities you see, and how do you explain them? (In particular, consider a number line.) How might you express the generalization?

6. In Case 14, students offer various models to illustrate the Commutative Property of Multiplication. Describe each model. Which models are convincing? Why? Which models seem limited? Why? What questions do these models raise for you?

7. What happens when you switch around the numbers in a division problem? What are the regularities you see, and how do you explain them? How might you express the generalization?

Planning for Student-Thinking Assignment

What task or question are you going to pose?

What mathematics do you want students to work on as they do the activity?

What questions might you ask while students are working?

Fourth Homework

Reading assignment: Casebook Chapter 4

Read Chapter 4 in the Casebook, "Expanding the Number System," including both the introductory text and Cases 15–19. As you read, think about your reaction to this question: What new ideas do students in each of these cases need to embrace as they expand their notion of *number*?

Write your statements about these ideas and bring them to the seminar. These statements will be shared with other participants.

Writing assignment:

Pose a question to your students related to the work of this seminar to examine how they think about these issues. Think strategically about how to create situations in which children will have a reason to make general statements about arithmetic and create models to show their thinking.

Then think about what happened. What did you expect? Were you surprised? What did you learn? Write up your question, how your students responded, and what you made of their responses (your expectations, your surprises, and what you learned). Include specific examples of student work or dialogue. Examining the work of just a few students, in depth, is very helpful.

At our next session, you will have the chance to share this writing with colleagues. Please bring three copies of your writing to the session.

R E A S O N I N G   A L G E B R A I C A L L Y   A B O U T   O P E R A T I O N S

Expanding the Number System

Mathematical themes:

■ In what situations are negative numbers useful?

■ What are the strengths and limitations of two models (number line and charge) for operating with negative numbers?

■ As the number system is extended to include 0 and negative numbers, how do we need to modify our ideas of what it means to order and to add numbers?

Session Agenda

Sharing student-thinking assignment	Small groups	30 minutes
Math activity: Numbers less than zero	Whole group Small groups	10 minutes 30 minutes
Math discussion: Which of two numbers is larger?	Whole group	30 minutes
Break		15 minutes
Case discussion: Beyond counting numbers	Small groups	30 minutes
DVD and case discussion: Beyond counting numbers	Whole group	30 minutes
Homework and exit cards	Whole group	5 minutes

Background Preparation

Read

■ the Casebook, Chapter 4

■ "Maxine's Journal" for Session 4

■ the agenda for Session 4

■ the Casebook: Chapter 8, Section 7

Work through

■ the Math Activity for Session 4 (p. 145)

■ the Focus Questions for Session 4 (p. 146)

■ Optional Problem Sheet 2 (p. 148)

Preview

■ the DVD segment for Session 4

Preview

■ Criteria for Representation-Based Proof

Materials

Duplicate

■ "Math Activity" for Session 4 (p. 145)

■ "Focus Questions" for Session 4 (p. 146)

■ "Fifth Homework" (p. 147)

■ "Optional Problem Sheet 2" (p. 148)

Obtain

■ graph paper

■ cubes or chips

■ index cards

■ DVD player

Using Two Models to Represent Addition of Positive and Negative Numbers

In the first two sessions of the RAO seminar, the domain of the numbers under consideration was primarily positive whole numbers. In Session 3, reversing the numbers in a subtraction problem introduced negative numbers. In Session 4, we further examine this extended domain: the set of integers {... ⁻4, ⁻3, ⁻2, ⁻1, 0, 1, 2, 3, ...}. Just as there are different ways to represent each of the operations with whole numbers, there are also different ways to model the set of integers. In Session 4 of RAO, while participants might call upon other models, two specific ways of modeling operations with integers are used extensively: the number line and the charge model.

The **number line** model for integers is an extension of the number line for whole numbers that was introduced in earlier sessions.

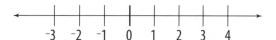

It is most common to orient the number line horizontally; however, a vertical number line is an appropriate match for the familiar thermometer, and this reference might be useful for participants.

The **charge model** uses chips or cubes of different colors. The name, *charge model*, is a reference to positive and negative electrical charges. One color represents positive units and another color represents negative units. For instance, positive 5 would be represented by 5 chips of a certain color and negative 3 would be represented by 3 chips of a different color.

When one positive unit joins with one negative unit, they cancel each other to create 0. Thus, each of the following arrangements is a representation of 0.

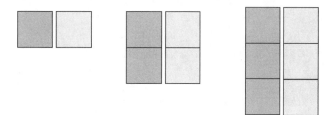

Consider using both models to show ⁻3 + 5.

One possibility on the number line is to begin at 0. Move 3 spaces to the left and then move 5 spaces to the right. Combining the two motions indicates the action of addition, and the sum is found by determining the final position on the number line, 2.

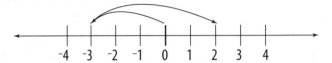

One possibility using the charge model is to gather 3 yellow cubes to designate ⁻3 and 5 green cubes to designate 5.

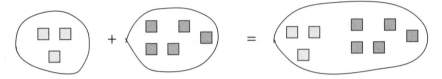

After addition has been modeled by joining the two sets, pairs of unlike cubes are matched. Each of these pairs is equal to 0 and can be removed, leaving 2 "positive" cubes representing the answer.

To summarize, each number is represented by the appropriate quantity of colored cubes. Joining the two groups of cubes represents addition. Pairs of different colors add to zero, and the answer is found by identifying the number and color of cubes remaining.

Maxine's Journal

October 22

In the previous session, the question about what happens when you switch the order of the numbers in a subtraction problem moved us into a new realm involving numbers less than 0, that is, negative numbers. If you have whole numbers, 0, 1, 2, 3 . . . , joined with their counterparts that are less than 0, ⁻1, ⁻2, ⁻3 . . . , you have the set of numbers called *integers*. In this session, we talked about adding and ordering integers. During the next session, we will work on subtracting integers, and in Session 6 we will touch on multiplying integers. | 5

As we work with integers, we need to think about the experiences of students as they expand their number system to include new kinds of numbers. There are actually two steps in this process: coming to recognize these new | 10 entities as numbers, and then figuring out how to calculate with these numbers. Some of the cases in Chapter 4 show students who have been thinking about counting numbers as they begin to think about 0 as a number. In other cases, students, like the participants in the seminar, begin to work with negative numbers. | 15

As I sat down to hear about the student-thinking assignments, it was clear that many of the participants had already moved into these issues with their students.

Sharing student-thinking assignment

At the end of the last session, participants met in small groups to plan their student-thinking assignment. I asked that they return to the same groups to | 20 share what happened.

Some participants decided they would ask their students what happens when they switch the numbers around in a subtraction problem and the participants were anxious to share the ideas that arose. Yanique, whose students are now very willing to voice their ideas, wrote about how one student worked | 25 with cubes to represent subtraction. If there are 8 cubes and 3 are taken away, there are 5 left: 8 − 3 = 5. If there are 3 cubes and you try to take away 8, the 3 are taken away, but 5 more are still needed. To represent that notion, the student wrote 3 − 8 = 00000. This student has an idea about a different kind of number and now needs to learn how it is written. Mishal commented that | 30 this student's thinking was similar to one student on the DVD clip we saw in Session 3. It was surprising to these participants that different students would come up with the same idea independently.

Kaneesha interviewed a student who, after some hard thinking, decided 3 –
7 = ⁻4. When asked to show where ⁻4 would appear on a number line, he drew

0 1 2 3 4 5 6 7
⁻4

putting the ⁻4 below and to the left of 0. When asked where ⁻3 would be, he
redrew the number line:

⁻4 ⁻3 ⁻2 ⁻1 0 1 2 3 4 5 6 7

Another group of participants explored the Commutative Property of
Multiplication with their students. They discussed whether 8 × 6 is the same
as 6 × 8 and said it confuses students because 8 rows of 6 chairs is, in fact,
different from 6 rows of 8 chairs. Charlotte pointed out that deciding whether
8 × 6 is the same as 6 × 8 depends on perspective. The person setting up the
chairs has to do the same amount of work either way. I thought that was a
good point and suggested that it is not so much whether 8 × 6 is the same as
6 × 8, but in what ways it is the same and in what ways it is different. The
equation 6 × 8 = 8 × 6 indicates "sameness" in a very particular way. In the
context of arrangements of chairs, the equation tells you that the total number
of chairs is the same for either configuration.

Math activity: Numbers less than zero

Whole group

I mentioned that those who had explored switch-arounds for subtraction with
their students had already moved into today's math activity. "It was already
clear from the last session that if you switch the numbers in a subtraction
problem, such as 4 – 3 and 3 – 4, you must begin to consider new kinds of
numbers. It was also clear that in order to think about these numbers, we need
new models and contexts to think about. Years ago, a second-grade teacher
shared something with me that came up when her students were curious
about turning the numbers around for subtraction. Her student explained that
if you have 9 fingers and take away 8, you have 1 finger left. However, if you
have 8 fingers up, and you try to take away 9, you are left with 0. Relying on
that familiar context that only considers whole numbers, the student decided
the answer is 0. We need to consider other contexts that allow us to talk about
negative numbers."

Together, participants came up with a list of contexts in which negative
numbers are used:
- Temperature
- Debits (having and owing money)
- Owing somebody something (other than money)
- Opposite direction (such as clockwise and counterclockwise)

- Latitude lines (the numbers on them)
- Time line (such as BC and AD)
- Sea level (above and below)

Temperatures below 0 and altitudes below sea level are conventionally indicated with negative numbers, and in some cases, owing money is, too. Phuong was the one who suggested clockwise and counterclockwise, thinking about positive and negative angles. This suggestion was outside the realm of many members of the seminar, so I offered some explanation. Latitude lines are marked north and south rather than positive and negative, but the behavior is the same. Dates, which are generally indicated as BC or AD rather than positive and negative, may be problematic since there is no year 0.

I told the group that these contexts would be very useful as we consider the behavior of integers, but I steered them away from using BC and AD. I explained, "If you try to figure out calculations on a model in which there is no 0, you are likely to become more confused."

I went on to say that, typically, there are two different models that help people think about integers. The first is a number line, which we had already been using in previous sessions. The other we call the "charge model." Because we have cubes at our convenience, we will use cubes of two different colors. Each cube stands for 1, but blue cubes (or some other color) will stand for ⁺1—one positive charge—and red cubes (or some other color) will be ⁻1—one negative charge. You can combine any number of red and blue cubes, but a red and a blue together is equivalent to 0.

Participants worked in small groups to experiment with the charge model. When I called them back together, I asked for one volunteer to demonstrate adding two positive numbers and another volunteer to demonstrate adding two negative numbers. Similarly, we looked at adding a positive and a negative, first with a positive result, then with a negative result.

Grace said, "That is cool. I can see why it turns out that adding a negative is the same as subtracting a positive."

When I asked her to explain, she held up the cubes to represent the problem we had been discussing.

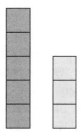

"Let's say you have 5 + ⁻3. Each of those 3 negative charges gets matched up with a positive charge to make 0, and you are left with 2 positive charges.

That is the same as if you started with 5 positive charges and just took away 3 of them."

I asked what happens when you have ⁻5 + 3.

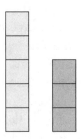

Grace continued, "Now you match the 3 positive charges against 3 negative charges, and those come out to 0, so you are left with ⁻2. That does not feel quite the same as starting with 3 and taking away 5, does it?"

Denise said, "But that sure looks like subtracting ⁻3 to me. If you start with ⁻5 and add 3, it looks the same as starting with ⁻5 and subtracting ⁻3."

Not very many people could appreciate Denise's insight, though their curiosity was piqued. However, I did not want to pursue it right then. In Session 5, we will have plenty of time to work on subtracting negative numbers. If we got into it now, we would not be able to address the issues that were on today's agenda. Besides, there were a few other ideas I wanted to address before the next session to help participants make better sense of what is happening when subtracting negative numbers. So I asked that we hold off on that idea until then. "Today we are going to stick with addition and deciding which of two numbers is greater."

I pointed out that in the last session, we looked at different models of multiplication and saw that different models offer different insights. For example, there are some things we see better with arrays and other things we see better with skip-counting on the number line. I suggested that here, too, we will find that sometimes the charge model works better and sometimes the number line is more appropriate. I would like the group to work with both models and decide which model seems to be better for which situations. Then, I distributed the "Math Activity" Sheet.

Small groups

The first problem asked participants to use each model to explore the Commutative Property of Addition over the integers. I looked over at M'Leah as she read the problem and saw her smile. "Using the charge model," she said, "it looks just like positive numbers. It is like students said on that DVD segment last time. You hold one stack of cubes in one hand and the other stack in the other hand, and then you switch them. Switching the order does not change the answer."

Phuong was using cut-outs on the number line again.

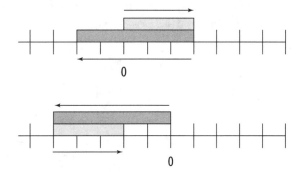

She explained, "First, I looked at 3 + ⁻5. I start at 0. The light gray bar shows my move of 3 units to the right and the dark gray bar shows my move of 5 units to the left. The dark gray bar extends 2 beyond the light gray bar, so I am at ⁻2. Then I flip both bars over and slide them down so that the right edge of the dark gray bar is at 0. I first move 5 to the left and then 3 to the right. I have the same amount remaining to the left of 0. So it has to be the same, no matter which order I add them."

M'Leah said, "Well, this is one way I think these models are different. It is a whole lot easier to see that order doesn't matter with the charge model."

Phuong said she still likes the aesthetics of the number line.

As I listened to the small-group discussions, I heard several participants state, "Integers are commutative," after they had confirmed that 3 + ⁻5 = ⁻5 + 3. Each time I heard this, I explained that commutativity is a property of an operation, not of a set of numbers. (When the participants thought of the property as one that involved a set of numbers, as opposed to an operation, it reminded me of Case 13, when Alice stated a rule about addition, and some students interpreted it to be simply the way the numbers worked—for any operation.) In Session 3, we demonstrated that addition is commutative over the whole numbers. Today we were checking to see that addition is also commutative over the integers. Now, when participants' attention was on integers—a new kind of number—it was as though the operation of addition receded into the background. I needed to highlight for them that we were discussing a property of *addition* and deciding whether it still holds for this new domain.

Kaneesha and Vera said that they just could not set aside this idea about subtraction. They saw what Denise was saying about taking ⁻5 and subtracting ⁻3. However, what if they started with ⁺5 and wanted to subtract ⁻3? How does that work with the charge model?

Even though I wanted Kaneesha and Vera to get to the other problems on the "Math Activity" Sheet, I decided to show them how to think about their question. "You know, there are times that mathematicians first make something more complicated in order to then make it simpler. One way to make it more complicated without changing the value is to add something that equals 0." I showed them that when you start with 5 and need to subtract ‾3, you will first add a form of 0—3 positive cubes and 3 negative cubes.

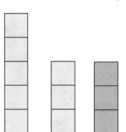

I paused here to make sure they agreed that the value of all these cubes together was still ⁺5. Once they concurred, I said, "OK, so now take away the ‾3."

A big smile spread over Vera's face, and Kaneesha said, "How cool. Now you have 8 positives left."

Next, I spent some time with Josephine to see how she was doing with the problems. Because she recently had new insights about the number line, I thought we might look at that together. When she said she knew that 7 degrees is warmer than ‾9 degrees, I suggested she draw a thermometer. We looked at where different temperatures are located on the thermometer, including 0 degrees, 7 degrees, and ‾9 degrees. Then I pointed out how the thermometer is similar to a number line, except the thermometer is drawn vertically.

Many of the groups were working on ideas about deciding which of two whole numbers is larger but did not get to the last question about comparing integers. We dealt with that as a whole group, so I will move to describing that discussion.

Math discussion: Which of two numbers is larger?

I started by asking how people know 9 is greater than 7. James pointed out that there are assumptions we make simply by stating that 9 is greater than 7. "Like, 7 feet is more than 9 inches," he explained. "When you say that 9 is greater than 7, you are assuming that you are using the same unit for both: 9 feet is longer than 7 feet, and 9 inches is longer than 7 inches."

I acknowledged that, yes, we are assuming the same unit, and then we proceeded to make a list.
- There is 7 in 9.

- Counting—9 comes after 7 when you count forward.
- 7 + 0 = 7, but 7 + 2 = 9. You have to add a positive number to get from 7 to 9.
- 10 − 9 = 1, but 10 − 7 = 3. If you start with the same amount, the more you take away, the less you have.

Expanding the Number System

- If you have a 7-ounce container and pour 9 ounces into it, it will overflow.
-

| | | | | | | | | | | ← More stuff

| | | | | | |

- 9 is to the right of 7 on the number line.
- 9 is farther from 0 on the number line.

I was quite impressed by this list. The lower-grade participants offered ideas that are present for their young students—that 7 is contained within 9, that 9 comes after 7 when you count. Participants also referred back to some of the generalizations we had worked on in Session 2. Comparing 7 + 0 and 7 + 2 is reminiscent of Lola's case, and comparing 10 − 9 and 10 − 7 goes back to Monica's case (taking a large or a small bite of bread).

Now I suggested we think about which is larger, 7 or ⁻9. "If we look at all these different ways to think about how we know 9 is larger than 7, do they help us decide about 7 and ⁻9?"

M'Leah said when she looks at the first item on the list, she does not really know how to think about whether ⁻9 is in 7 or if 7 is in ⁻9. Also, looking at the second item, if you start counting forward from 0, you will never get to ⁻9.

I acknowledged that there are several ways of thinking about 7 and 9 that are restricted to the counting numbers or positive quantities. "It does not make sense to try to pour ⁻9 ounces into a 7-ounce container." I wrote up headings for three different categories: "supports the idea that 7 > ⁻9," "supports the idea that ⁻9 > 7," and "does not apply to negative numbers." Then I said, "M'Leah pointed out that some items on our list would fit into this third category. Let's look at the items that do allow us to think about negative numbers. You might also look at the list of contexts to see if that helps."

Josephine pointed out that 7 degrees is warmer than ⁻9 degrees. That means that 7 is larger than ⁻9.

M'Leah said she would rather have 9 candies than have to give 7 candies away.

Grace said someone with $7 is richer than someone who owes $9.

Mishal countered, "If someone has $7 and owes $9, that person's debt is more than what he has. That sounds to me like ⁻9 is more than 7."

Kaneesha, who had been so enthralled by the charge model for integers, held up 7 blue cubes and 9 red cubes. "Look at this. The red stack has more stuff. Doesn't that mean ⁻9 is larger than 7?"

Antonia said that if you add 16 to ⁻9, you get to 7. "Because you add a positive amount to ⁻9, that must mean that 7 is larger."

Leeann said, "But, if you look at the number line, ⁻9 is farther away from 0."

Yanique countered, "And, if you look at the number line, 7 is to the right of ⁻9."

Jorge said, "It depends on how you define bigger. If bigger is farther from 0, then ⁻9 is larger. If bigger is warmer temperature or having, instead of owing, or movement to the right, then 7 is larger."

I used Jorge's insight to move the discussion in that direction. "Jorge is absolutely right. We see that if we look at it from one perspective, we conclude that ⁻9 is bigger than 7, and if we look at it a different way, we conclude that 7 is bigger than ⁻9. As a group, we have come up with several arguments for both conclusions. Kaneesha's cubes look like ⁻9 is larger than 7; Leeann pointed out that ⁻9 is farther away from 0 than 7; and Mishal pointed out that a debt of $9 is more than a savings of $7. Those ways of looking at it suggest ⁻9 is larger than 7. However, if we look at it from another perspective, we conclude that 7 is bigger than ⁻9. As Yanique said, 7 is to the right of ⁻9, and Antonia pointed out that you add a positive amount to ⁻9 to get to 7. So Jorge's right, it depends on how you define bigger."

I paused for a moment to see if there were any comments. When there were none, I continued. "When we expand the number system to integers, we have certain properties that are important. It is our goal to expand the number system in such a way that we preserve these properties. Some of the rules we have been looking at in the last few sessions are important to remember. For example, in your small groups, you recognized that integers preserve the Commutative Law of Addition. For another example, remember the rule that came out of the Kindergarten case?" I wrote up, If $a < b$, then $a + c < b + c$. "So, let's say you start with ⁻9 and 7, and you add 10 to each."

I gave people a few minutes in small groups to go back to the inequality we had discussed two sessions ago and replace the variables with the numbers ⁻9, 7, and 10. After a few minutes, we came back together to discuss their observations. Lorraine said, "If you look at that rule, you have to say $7 < ⁻9$. Because ⁻9 + 10 = 1 and 7 + 10 = 17, and because 1 is less than 17, ⁻9 must be less than 7."

Charlotte said, "I see that. But why do we care about that rule and not that the number farther from 0 is larger?"

I answered, "Mathematicians have been pretty good at figuring out which decisions to make to create the greatest consistency in the system. So, maybe we just need to accept this one. By definition, if you add a positive amount to one number to get a second number, the second number is larger. So 7 is larger than ⁻9."

Mishal said, "That still seems backward to me. If I owe $9 and have $7, I owe more than I have. It doesn't make sense."

Charlotte suggested, "You might think about having or owing money in terms of net worth. If person A has $7 and person B owes $9, then you say person A's bank account is worth more than person B's."

I acknowledged that Charlotte's idea is consistent with the definition of which number is bigger. "However, that is not the end of the story," I said.

"Sometimes it is useful to compare ⁻9 and 7 in the way that Kaneesha, Leeann, and Mishal have been doing—that is, to compare how far away from 0 they are. When we want to think about numbers in that way, we use a different term and different notation. We say that the *absolute value* of ⁻9 is greater than the *absolute value* of 7." I went back to the heading I had written toward the beginning of the discussion and now inserted the correct notation: "supports the idea that $|{-9}| > |7|$." "We say that the absolute value of ⁻9 is equal to 9 because ⁻9 is 9 units away from 0." I wrote: $|{-9}| = 9$.

There were some "ohhhh's" and comments in the room. "That is what absolute value means." "I never really understood that notation."

After a pause, I went on. I wanted to make sure everyone now understood how to answer the question of which is larger, so I asked, "Which is larger, ⁻9 or ⁻7?"

Risa said, "⁻7 has to be larger because it is to the right on the number line. But the absolute value of ⁻9 is larger than the absolute value of ⁻7."

Mishal added, "You know, I can accept that. It would have driven me crazy to say that ⁻7 is larger than ⁻9 if I did not have that idea of absolute value."

Case discussion: Beyond counting numbers

To transition our thinking to the cases while I distributed the Focus Questions, I said, "We have been working to extend our own sense of numbers to learn how to work with numbers less than 0. It is quite a stretch, isn't it? In the cases you read for today, students are also extending their view of the number system. We first came to know numbers through counting: 1, 2, 3, 4, 5. And so, when we want students to be thinking about other numbers—numbers you do not normally use when counting, such as 0, fractions, and negative numbers—there is a lot of stretching to do. The cases in this chapter are about students confronting 0 as a number and then thinking about negative numbers. Let's take a look at how students work to make sense of these numbers."

At the first table I visited, James was saying, "I really do not get why 0 is such a difficult thing."

Mishal said, "Well, it's like in the case about cats and dogs. It feels strange to be talking about cats and dogs when you have nothing representing dogs."

Yanique said, "It is hard to have something tangible to represent nothing. 'Nothing' is hard to represent in a concrete way. The moment you use something to represent 0, you have something."

Claudette said, "But look at what students did in Cassie's case. They thought of 0 as holding a position next to the 12 cubes that stood for 12 cats."

Mishal acknowledged, "That is really interesting that first graders thought of that."

At another table, M'Leah explained, "When my students are working on ways to make 7 and tell me that 7 green cubes and no yellow cubes is a solution—and some even say 7 + 0—then I know they are thinking about 0 as a number."

Claudette said, "I could relate to the case about 12 cats and 0 dogs. I know so many students, like Thomas who will not accept a solution with 0. He just says, 'It looks funny.' "

Denise said, "It's not just young students who have trouble. The fifth graders were also struggling with how to think about multiplying by 0."

Risa added, "It doesn't stop at fifth grade. Have we ever thought about it before? I had to work hard to think about the stories those fifth graders came up with."

At the next table, the group was also talking about 0. Lorraine said, "I am thinking about how my students sometimes have trouble learning to use a ruler. They think 1 should be at the end of the ruler because that is where you start. They are not thinking that 0 is really the starting place, and you have to move one unit over to get to 1."

DVD and case discussion: Beyond counting numbers

I was surprised that the groups spent so much time talking about 0 and hardly started on the cases about negative numbers. Regardless, before we started the whole-group discussion, I showed the DVD segment of third graders working on word problems about temperature using values above and below 0. The whole-group discussion would focus on negative numbers.

After I turned off the DVD player and before I could even ask a question, Leeann asked, "Why do the positive numbers have to be to the right?"

Phuong said, "They don't. Like what we just saw, on a thermometer, the positive numbers are on top."

Antonia was intrigued by the option of using vertical or horizontal number lines. She said, "I know the ideas are the same in each, but they feel different."

Denise said, "The girl on the DVD put ⁻4 on top, and I wanted to say, 'No!' However, I liked the way the teacher handled that. She did not say the girl did it wrong. But, did she?"

I answered this question directly. "There is not anything conceptually wrong with putting negative numbers on top or to the right. It is merely a convention to put positive numbers to the right or on top."

Grace said, "You know, whenever I draw a number line to help me think about a calculation, I always put the larger numbers to the left. I always have to turn things around in my head when I look at someone else's number line."

Charlotte said, "I do the same thing."

I commented, "When you are drawing a number line just for yourself, it does not matter which direction you use. The issue is really when you want to communicate with other people—either to understand their ideas or tell them how you are thinking. Because the convention is to put larger numbers to the right, it seems easier to build your number line that way from the beginning."

Denise asked, "So what about the girl on the DVD? Should the teacher tell her to put higher numbers on top because that is the convention? It would make things easier for her in the long run."

Quite a few participants were adamant when they said, "No!" Madelyn explained, "If there is not anything conceptually wrong, it might be more confusing for the student to be told that she is wrong. I bet she will get used to looking at thermometers that have numbers on top. If her classmates and teachers all show their number lines the same way, I bet she will eventually turn it over."

I asked Grace and Charlotte about their experience. Both said that they did not learn to use number lines when they were students; number lines were just images they had in their heads. Maybe if they had been shown the conventions when they were young, and if they used them regularly that way, then they would make their number lines using the standard convention.

Grace started shaking her head, and pointing to the number line that was on the board, said, "There is something else that bothers me. It is about the space between $^-1$ and 0 and through to a $^+1$. I can see it on the number line, and I can reason it out, but it still bothers me that this space is there."

I was not really sure what Grace was talking about, so I asked her if it bothers her when she looks at a thermometer.

Grace responded, "No. I can feel temperature. That does not bother me so much."

Madelyn pointed out that the thermometer is different because the temperature we call 0 is arbitrary. I checked to make sure she was thinking about Fahrenheit and Celsius scales and then elaborated for the group. "When the temperature is 0 degrees Fahrenheit, it is actually colder than when it is 0 degrees Celsius." Then I turned backed to Grace. "Is it easier to think about with temperature because there really is no such a thing as no temperature? Zero is just a place on the scale that is not so different from $^-5$ or $^+5$?"

Grace nodded, so I went on. "Can you see $^-\frac{1}{2}$ and $^-\frac{1}{4}$ on the number line? Do you see that the space between $^-1$ and 0 gets filled up with numbers like those?"

Charlotte joined in, "So you are saying there is this place where the positives and negatives meet. If I went from $^-\frac{1}{4}$ to $\frac{1}{4}$, at some moment I would be at 0."

I was still trying to understand what the issue was, and so I tried again. "Let's keep looking at the number line but thinking about some of our contexts—such as having something or owing something. When we are at $^-1$, we have to give one piece of candy; when we are at $^-\frac{1}{2}$, we have to give a $\frac{1}{2}$ piece of candy. As we get closer to 0, we have to give away less and less, but we still owe. Then, when we get to 0, suddenly we do not owe any more, and when we continue to move to the right, now we have candy. So as we pass through 0, there is this change of state. Something significant happens there."

Grace said, "It's like you can always cut up the cookie, until there is just dust. You get closer to 0, but you don't go below it."

Lorraine added, "I think it's hard to visualize 0. I can think of *needing* something, which is negative, and I can think of *having* something, which is positive, but it's hard to think about having 0."

I suggested we look back at the contexts we came up with at the beginning of the session. Which of those are like temperature, where it is easy to think about going from positive, through 0, to negative, and which of those seemed to indicate a change of state? The group agreed that latitude lines, time lines, and sea level were more like temperature. However, the change in direction was more like having and owing money. The idea of using signs to indicate movement in clockwise or counterclockwise directions was so new to everyone but Phuong that they could not relate it either way.

As the session ended, I told the group that the cases they would be reading for homework address some new ideas, and we would use those ideas to think further about negative numbers in the next session.

Exit cards

In their exit cards, I asked people to comment on the ideas about negative numbers they are taking from this session and what questions they still have about negative numbers.

Josephine wrote about her new understanding of the number line. "I had never learned about it from any of my teachers. Often times, one's struggles are the result of how they were taught. Fortunately, it is never too late to learn. I am becoming so confident about math that I want to learn more."

Some people, such as Kaneesha and June, felt pretty excited by having the two representations of negative numbers—the number line and the charge model.

James commented that he liked discussing the different models. He said that he came to a realization when we were working on the question of which is larger, ⁻9 or 7. "I don't know why it never occurred to me that the two interpretations of *larger* could co-exist, even though they were two ideas I had thought about. I kept trying to cram both ideas into one model."

Even with the two representations, some people are feeling uneasy about how different negative numbers are from the numbers they are familiar with. June is beginning to think about subtracting negatives and questions her idea that "−" means moving toward the left on the number line and "+" means moving toward the right. "In the problem ⁻3 − (⁻2), I am not sure if that applies

395

400

405

410

415

420

425

430

because I am confused about whether the answer is ⁻5 or ⁻1." This is exactly what we will be working on in the next session.

Charlotte is also moving into next session's territory. "I become nervous thinking about subtracting a negative from a negative or multiplying a negative times a negative. I think it is the spookiness of working with one thing and another thing when neither really exist but are only ideas." 435

M'Leah asked, "Is it all right to continue creating context to understand the meaning? I am not convinced that I understand the concepts fully without creating a context in which they will fit. I will work on it, though." 440

I plan to address some of these questions in my responses to their homework.

Responding to the fourth homework

October 25

I found this set of homework to be fascinating because of what I learned from it about student thinking. I was also impressed by the progress I saw from the last student-thinking assignment to this one. Mishal has figured out what it means to write a case and Miriam has slowed down to start to hear her students' ideas, which are very strong. Yanique's students have learned that mathematics can be very interesting and are glad to share their ideas. 445

M'Leah's student-thinking assignment illustrates first graders tackling questions about subtracting a larger number from a smaller one. 450

M'Leah

Negative Numbers in a First-Grade Classroom

I teach in a first-grade, English-speaking, two-way bilingual classroom. My students learn in two languages—Spanish and English. Although I see my students for only half of the school year, I enjoy learning how they interpret math in the two languages. 455

This year, one of the two classes that I teach seems to be understanding and verbalizing math ideas more than I have ever seen in the past. Some students in the class seem to have a deeper understanding of the typical math problems that we have been working on. Other students, however, are still struggling with some concepts. 460

After sitting through the DMI sessions and reading the cases, I wondered how some of my students might think about the idea of negative numbers. I was not sure how my students would view the problems when they were written and when they were presented in context. Because some of my students are still struggling with the idea of combining two numbers, I decided to work with a small group of five students on the new ideas about negative numbers. 465

The five students that I selected are each a different type of learner. Two students had been retained this past year due to their reading scores, not their math scores. Therefore, the math we have been working on, up until this point, has been more of a review for them. The other three students have been very quick to pick up new concepts. Therefore, I chose these five students to approach the problems. 470

I first wrote down, 5 – 3 = ? I gave students unifix cubes and asked students to read the problem and then solve it. All students said, "5 take away 3 equals . . ." They were quick to respond with "2." They drew a picture to match the problem on their pieces of paper. 475

Jemea drew:

1   2

Alejandro drew: 480

2

Manuel drew:

2

Efrain drew:

Consuela drew:

1   2

I felt pretty confident that they had an understanding of the problem. I then asked them to put the problem into context, which they all did successfully. 485

Then, I wrote the problem 3 – 5 = ?

Students looked at the problem with many questions.

ALEJANDRO: Hmmm . . . I am not sure about this one.

I did not know how students would read the problem, so I asked them to put the problem into context. I anticipated students reversing the numbers to have it make sense to them (5 – 3 = ? as opposed to 3 – 5 = ?). 490

ALEJANDRO: Well, if you need 5 sandwiches from me but I only have 3 then . . . well, it would be 0. None! 3 – 5 = 0. You are going to take away 3 when there are only 3 so that will leave none.

Expanding the Number System

He seemed pretty convinced that 0 was the answer. Manuel and Consuela, at first, agreed with Alejandro.   495

MANUEL AND CONSUELA: The answer is nothing. There is nothing left. No sandwiches.

Jemea seemed to have another idea.

JEMEA: (Holding up 3 fingers) If you start with 5 and take 3 away   500
that means that there are 2 more than you need.

Jemea seemed to have a sense that there was something about the 2 that was important. This sparked an idea in Manuel's mind.

MANUEL: Mrs. Panos, you need 5 sandwiches from me, but I have only 3 sandwiches.   505

TEACHER: Yes.

MANUEL: That means that I need to go down to the lunchroom and get 2 more sandwiches from Millie, or I can take 2 sandwiches from someone who owes me 2 sandwiches.

Manuel seemed pretty confident about his answer. He realized that 2 sandwiches   510
were owed to someone; therefore, he looked for 2 more to give me.

Consuela looked at Manuel with confusion.

CONSUELA: No, you need 6 sandwiches because 5 is lower than 6.

I was interested in what Consuela was thinking. She seemed to be rather quiet through the conversation, yet something about what Manuel was saying   515
sparked her interest.

CONSUELA: Your problem says 3 take away 5, right?

TEACHER: Yes.

CONSUELA: Well, if you took 5 away from 3 you would have nothing left. That is why we need 6 sandwiches. Then we will   520
have enough to give you.

Consuela was changing the problem so that it made sense to her. After she made that point to the group, she seemed to "tune out" for the rest of the lesson. She was very satisfied with her response.

Alejandro then chimed in.   525

ALEJANDRO: But guys, look. What if we had 2 sandwiches and needed 3 more? Then you would have 2 + 3 = 5. So that means that you, Mrs. Panos, will have to start with 2 sandwiches.

Alejandro was presenting his ideas to me, and so I asked him to explain what   530
he meant to the group.

ALEJANDRO: *If Mrs. Panos started with 2 sandwiches, then if we gave her our 3 sandwiches, then she would end up with 5. She would be happy.*

MANUEL: *OK, but if I had 3 sandwiches and I need to give Mrs. Panos 5 sandwiches, I would just go home and make 2 more to give her.* 535

The bell rang, as it was time for lunch. I tried to squeeze in one last point. I asked students to make a representation of the math that we were just doing. Most students drew some sort of 3 and 2 split to make 5. I asked them how 540 *the two problems connected but nobody could explain it. Besides, lunch was waiting for them. Yet, as students walked away, Alejandro stopped and looked at me with his paper in his hand.*

ALEJANDRO: *Mrs. Panos, there are 3 in the top problem and 3 in the bottom problem just like there are 2 in the top problem and* 545 *2 in the bottom problem.*

What Alejandro saw:

5 − 3 = 2

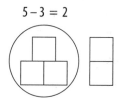

*noticed: "3 on top and 3 on bottom.
2 on top and 2 on bottom."*

3 − 5 = ?

Class was over; therefore, so was our discussion. I would love to revisit this 550 *problem later in the year with the rest of the class.*

I found this discussion among first graders fascinating. They were touching on ideas of negative numbers, but I also have the sense that Alejandro is noticing important relationships between addition and subtraction. In my response, I wanted to discuss students' ideas about 3 − 5 and also ask 555 M'Leah to think about how this case is related to the content we will be working on in Session 5.

Dear M'Leah,

What a fascinating discussion you had with this group of five first graders. How interesting that they could take on a context for thinking about 3 − 5 and 560 *go beyond "it is impossible" or "you end up with 0." Presumably, without ever having heard of negative numbers before, four of your students could think about this situation.*

Expanding the Number System

Alejandro could create a story problem for 3 − 5: You need 5 sandwiches, and I have 3. I assume he came to that situation by thinking about what it would mean to have 5 − 3: You need 3 sandwiches and I have 5, so I could give you the 3 and I will still have 2. In the 3 − 5 situation, I can give you the 3 that I have and then something else needs to happen so that you could get your 5; 2 sandwiches need to come from somewhere else. ⁵⁶⁵

*Some participants may have thought that if they introduced a problem that involved 3 − 5, they should show their students the answer, ⁻2. I agree with your judgment that it would not help your students to show them the notation for negative numbers at this point. It is enough that they have the ideas they have expressed—that you could have a story context that fits 3 − 5, and the situation involves doing something with 2, but it is **not** that 3 − 5 = 2. The 2 sandwiches still needed involve something different from having 2 sandwiches left over.* ⁵⁷⁰ ⁵⁷⁵

There is something else there that Alejandro understands. You can alter the problem so that it is illustrated by 2 + 3 = 5. If Alejandro has 3 sandwiches and gets another 2 from somewhere else, then there are 5 sandwiches to give to Mrs. Panos. Is this the idea you were thinking about in your exit card? ⁵⁸⁰

By the time you read this, you will have read the cases for our next class, some of which have to do with how addition and subtraction are related. We will be looking at a DVD of second graders talking about this idea, and we will consider this idea as we continue our investigation of negative numbers. My sense is that Alejandro is very clear about some of these connections. I wonder about the other students in the discussion. ⁵⁸⁵

Regarding the issue about the relationship between addition and subtraction, I am trying to figure out how (and when) first graders can think about it and how (and when) second graders can think about it. If students notice that some students solve a problem by subtracting and others solve that same problem by finding a missing addend, how do they make sense of that? When can they take on the question of whether that will always work? If it is of any interest to you and your students, I would love it if you were to explore that question with them for your next assignment. ⁵⁹⁰ ⁵⁹⁵

Also, in your exit card, you wrote, "I am not convinced that I understand the concepts fully without creating a context in which they will fit." I think that is exactly the right impulse—to create a context to make sense of the ideas—at least for the content we are working on. At the same time, some of the decisions that get made about how the system works are determined by keeping the system consistent, not by thinking about contexts. If we come to one of those decisions that are determined by consistency, I think it is a good idea to go back to see if there are contexts that can fit, to give it more meaning. This problem might come up in Session 5. If so, let's talk about it. ⁶⁰⁰

Maxine ⁶⁰⁵

Detailed Agenda

Sharing student-thinking assignment (30 minutes)

Small groups

The first activity of this session provides the opportunity for participants to read and discuss their writings about the thinking of their students. In the last session, participants met in small groups to plan for this assignment. Participants should reassemble in the same small groups so they may share their experiences. Suggest they read their colleagues' examples of student thinking with an eye toward questions about teachers' moves and ask the following questions: "Once they elicited student ideas, what did they do next?" "What tools did students need to move forward and what did the teacher provide?" Let the group know that they will be looking at some of the cases in Chapter 4 to examine the actions of the teachers. Remind them to read the set of papers before beginning any discussion; their conversation should include similarities and differences among the papers. Let them know they will have 30 minutes for this activity.

Defining Sets of Numbers

The numbers 1, 2, 3, 4, . . . are referred to as the *counting numbers* or *natural numbers*. If 0 is included, the set of numbers is called the *whole numbers*. In this session, we will consider what happens when we enlarge our number system to include not only 0, but also ‾1, ‾2, ‾3, ‾4, . . . This enlarged system is referred to as the *integers*. In Session 2, participants examined numbers such as $\frac{3}{4}$ and $\frac{9}{8}$. Numbers that are formed by dividing two integers are called *rational*. The set of all rational numbers includes the integers, which can be written with a denominator of 1, as well as what are commonly known as fractions and decimals. In order to completely fill the number line, we also need to consider irrational numbers. These numbers are beyond the scope of the RAO seminar. They include numbers such as π and $\sqrt{2}$. The set of all rational and irrational numbers is called the *real* numbers, which comprise all the numbers on the number line.

Math activity: Numbers less than zero (40 minutes)

Whole group (10 minutes)
Small groups (30 minutes)

The math activity includes two components: 1) a whole-group discussion to elicit situations in which negative numbers are useful and to introduce two

Expanding the Number System

ways of representing negative numbers and 2) small-group work on ordering and adding integers using these two models.

Let participants know the math topic for this session is negative numbers and then clarify the terms *whole numbers* and *integers*.

Begin the discussion by brainstorming a list of situations in which negative numbers are used. Give participants a minute or two to write down their ideas and then compile a list on the board or on easel paper. See "Maxine's Journal" (pp. 122–123) for an example from one seminar.

Be sure that participants recognize that while many people understand negative numbers, many also have difficulty conceptualizing the set of negative numbers as part of a consistent number system. Explain that there are some abstract models that have been designed to help people make sense of negative numbers. Demonstrate two of these models, the number line and the charge model. Encourage participants to use these models as tools for their thinking about positive and negative numbers.

Once you have demonstrated the two models, ask the group to think about how to use the models to represent addition. Participants should consider the problem ⁻3 + 5 using both models. Give them a few minutes to think about the problem, perhaps suggesting they talk to the person sitting next to them to figure out what it would look like in both models. Then ask for someone to share the number-line strategy. Once the number-line strategy has been demonstrated, ask for a volunteer to show how he or she used the charge model to represent the same addition problem.

Let participants know these models are more abstract than the situations they had generated and that this activity is designed to help them become familiar with using these models. Remind participants that, as they saw with the models for multiplication, different models highlight different ideas. They might find that it is easier to visualize some aspects of the number system using the first model and other aspects using the second. It is important to become flexible with using both models in a variety of situations and to analyze the benefits of each model. Distribute the "Math activity" and, suggest that as they work on the questions they consider, "What is this model helping me see?" and "What are the limitations of this model?"

As you organize participants into small groups tell them that, in this session, they will be exploring two aspects of integers with these models: addition and ordering (considering which of two integers is the greater). Let them know subtraction of negative numbers (which many might remember with emotion) will be examined in the next session, and this work on addition and ordering numbers will be the basis for their upcoming work on subtraction.

As you circulate among the small groups, listen carefully to participants and help them express their observations correctly. For instance, when noticing that 4 + (⁻6) = (⁻6) + 4, some participants might state, "Negative numbers are commutative." Point out the Commutative Property is a property of an operation—in this case, addition. That is, $a + b = b + a$ holds even when a or b is a negative number. Assist them in rephrasing their observation as the Commutative Property of Addition holds when the numbers are integers and not just whole numbers. (In fact, addition is commutative for all real numbers; however, in this seminar, only rational numbers are examined.)

See "Maxine's Journal" (pp. 122–126) for more information on other ideas that these discussions might include. This information is particularly important because the whole-group discussion will be focused on the question of order. Also, once you have the sense that participants understand the two models, invite them to analyze what each model offers and how a model might be a source of confusion. One example might be that with the charge model, ⁻8 is represented with more cubes than 3, even though ⁻8 is smaller than 3.

Math discussion: Which of two numbers is larger?   (30 minutes)

Whole group

Focus the math discussion on the question of how to determine which of two integers is larger. When working with whole numbers, people develop many intuitive ways to think about which of two numbers is larger. However, when comparing negative numbers—or a negative and a positive—these tests for which number is larger are likely to be inconsistent. In this discussion, participants confront this inconsistency in order to identify the need for a definition of which of two numbers is the greater.

Participants' work on Questions 2–4 from the Math activity will be the basis for this discussion. Begin by establishing a variety of ways to determine which of two positive numbers is larger; then examine what happens if those strategies are used to determine which is larger, ⁻9 or 7. The strategies should include the number line and the charge model. When the strategies for determining order are seemingly inconsistent, use that opportunity to define absolute value and then define order using the number line.

During the discussion, several points are likely to arise. Here are just a few of these ideas:

- On the number line, 7 is to the right of ⁻9 and so 7 would be larger.
- The charge model seems to support the notion that ⁻9 is larger because more cubes are used to represent ⁻9 than to represent 7.
- The idea that ⁻9 is farther from zero supports the notion that ⁻9 is larger.
- If the "what to add on" strategy is applied, then 7 is larger because you need to add a positive amount (+ 16) to ⁻9 to get 7.

Consult "Maxine's Journal" (pp. 126–129) for an example of such a discussion.

In continuing the discussion, point out that the set of statements that apply in a consistent fashion when the two numbers are positive will lead to different conclusions when the numbers to be considered include negative numbers. The confusion about what it means to order numbers can be resolved by recognizing that there are different relationships being considered.

Clarify that, by definition, 7 is greater than ⁻9 let participants know that to capture the sense in which ⁻9 is larger than 7, we define a new term—*absolute value*—and introduce the notation $|7|$ and $|{-9}|$. This notation is read as "the absolute value of 7" and "the absolute value of ⁻9." Define *absolute value* as "the distance a number is from zero." Distance is always a positive quantity or 0, so the absolute value of any number is always greater than or equal to 0. Because both 9 and ⁻9 are 9 units away from 0, both numbers have an absolute value, 9. In notation, $|9| = 9$ and $|{-9}| = 9$, or $|9| = |{-9}|$.

After defining absolute value, ask participants what questions this symbol brings up. Once the group has had a chance to resolve any questions, ask them to work on Question 4 concerning a rule about ordering numbers that applies consistently to all pairs of numbers, no matter if they are both positive, one positive and one negative, or both negative.

These two statements should be listed:
- On a number line, the number on the left is smaller. (Or the number to the right is greater.)
- In order to make two different numbers equal, a positive amount must be added to one of the numbers. The number you add this positive amount to is the smaller number.

Break   (15 minutes)

Case discussion: Beyond counting numbers   (30 minutes)

Small groups

This chapter explores student thinking as they work to expand their idea of the number system beyond the counting numbers. The first three cases examine the thinking necessary to recognize zero as a number, and the last two cases extend the idea of number to include negative integers. In these cases, not only are ideas of what it means to call something a number being challenged, but also once new numbers are accepted, it is also necessary to revisit generalizations previously made. The module "Making Meaning for Operations" provides examples of this as students incorporate fractions into their number system. The cases in Chapter 4 examine the rethinking necessary when including zero and negative integers.

Distribute the Focus Questions for Chapter 4 and have participants work in small groups to answer the questions.

Focus Question 1 explores the thinking of first graders as they work to make sense of zero as a number.

Focus Question 2 examines fifth graders as they consider what it means to operate with zero. Participants explore the specifics from the cases and are asked to consider what other ideas about zero they would want their students to develop.

As they consider Focus Question 3, participants examine young students' beginning ideas about negative integers. You can also encourage participants to compare the thinking of students with their own work on these ideas.

Focus Question 4 is based on the same case but invites participants to analyze the actions and statements of the teacher. What are the teacher's moves that support and encourage student discussion of these ideas?

Finally, Question 5 explores the Commutative Property of Addition and how this property must be revisited when the numbers involved include negative integers. Participants should first examine the logic of the arguments presented in the case, "How does this reasoning apply to *all* numbers?" Participants should also create their own arguments.

DVD and case discussion: Beyond counting numbers

(30 minutes)

Whole group

Show the DVD of third-grade students working with temperature problems before the whole-group case discussion. Suggest that as participants watch the DVD, they keep in mind the ways students are making sense of negative numbers and the role of zero.

After the DVD, begin the whole-group discussion by soliciting some comments about students' thinking about zero. Remind participants that they may comment on the DVD segment or on the cases. After discussing their comments on students' thinking, turn the discussion to Focus Question 4 and solicit examples of teachers' moves from the print and DVD cases that support such student thinking.

Homework and exit cards

(5 minutes)

Whole group

Distribute the "Fifth Homework" Sheet.

Exit-card questions:
- What ideas about negative numbers are clear to you and what ideas are you still wondering about?
- What was this session like for you as a learner?

Expanding the Number System

Before the next session . . .

In preparation for the next session, read the participants' cases and write a response to each one. See the section in "Maxine's Journal" on responding to the fourth homework. Make copies of both the papers and your response for your files before returning the work.

DVD Summary

Session 4: Adding Integers

Second-grade class with teacher Danielle (7 minutes)

The clip begins with slides of the problem students will be discussing:

> Last weekend I was planning to take my dog for a walk. Early in the morning when I got up, it was ⁻4 degrees F. I decided to wait a few hours and went out later in the morning when it was 12 degrees F. How much warmer was it by the time I went for my walk?

One student draws a vertical line on the board and labels ⁻4 near the top. She says she will minus 4 and get to 0. She draws a loop from ⁻4 and labels it 4, marking 0 as a point below ⁻4. She continues, "Then I plussed 12." She draws another loop, this one from the point she marked as 0 to a point below 0 on her line indicating the loop is a jump of 12. Then she says she "plussed the 4 and the 12 and got 16 as an answer."

Danielle, the teacher, invites students to compare their work to this representation and determine how their work is different from this drawing.

One student indicates that she had switched the ⁻4 and the 12. Danielle repeats what this student has said and asks if it matters in solving the math problem which way you draw the points on the line. The students explain that the answer would be the same either way.

Danielle draws a vertical number line and places the 12 near the top and the ⁻4 near the bottom and then rephrases the argument that the first loop shows a difference of 12 and the second loop shows a difference of 4 and that $12 + 4$ is still 16.

Danielle asks students to compare the two drawings by asking what is the same. Students say they see a jump of 4 on both and a jump of 12 on both.

Danielle indicates that she sees $12 + 4$ on one and $4 + 12$ on the other and asks students what they think about that. One student says the total is still the same.

A new problem is posed:

> Last December I visited my grandparents in Canada when the temperature was 25 degrees F. Five years ago when I visited them, the temperature was 32 degrees colder than that. What was the temperature five years ago at my grandparents?

A student suggests making a number line, and Danielle draws a horizontal line and labels a point, 25. The student suggests subtracting 25 to get to 0, and Danielle acknowledges this is a good strategy.

Danielle asks students to show with their hands where 25 degrees colder would be and most students show moving to the left.

The student continues with the strategy of subtracting 25 to get to 0. Then he suggests subtracting 5 more.

Danielle asks where they will be when they subtract 5 from zero and the class says negative 5.

The student finishes the problem by subtracting 2, indicating he ends at negative 7.

The clip ends as Danielle says that was awesome thinking.

REASONING ALGEBRAICALLY ABOUT OPERATIONS

Math Activity

A note about terminology: The numbers 1, 2, 3, 4, . . . are referred to as the *counting numbers* or *natural numbers*. In this session, we will consider what happens when we enlarge our number system to include 0, ⁻1, ⁻2, ⁻3, ⁻4, . . . This enlarged system is referred to as the *integers*.

1. Choose one of the models for the integers that have been introduced (the number line or the charge model) and use it to explore "switch-arounds" (or the Commutative Property) for addition among integers. Start by considering specific pairs of integers, asking, for instance, whether ⁻7 + 3 is equal to 3 + ⁻7. Then extend your reasoning to *all* pairs of integers. Be sure your justifications apply to all possibilities, including the addition of 0, negative numbers, and positive numbers. When you feel satisfied with your exploration of the Commutative Property of Addition using one model, check it out with the other model.

2. List all the ways you know that 9 is greater than 7.

3. Now consider how to compare integers. Using the same two models (the number line and the charge model), how would you compare two numbers (positive with positive, negative with negative, and negative with positive) to determine which is the greater number?

4. Go back to your list of the ways you know that 9 is greater than 7 (Problem 2). Given your response for Problem 3, which of those methods still apply when both numbers are negative or when one is negative and one is positive? Can you state a single rule for order that applies for all integers?

Focus Questions

As you respond to the focus questions, refer to the ideas you noted as you read the cases.

1. In Cases 15 and 16, first-grade students are working on ideas about zero. What are their questions about zero? What characteristics of zero do they understand? What other ideas about zero would you want them to develop?

2. In Case 17, we see older students grappling with a similar question. What is their question? How do they work on it? What conclusions are they able to make? What other ideas about zero would you want them to develop?

3. In Case 18, a first-and-second-grade class is beginning to consider that there might be numbers on the number line to the left of zero. For each passage, identify the math idea that is being discussed:
   - Lines 236 to 259, ending as the teacher asked, "What does anybody think about that spot that Carl and Seth just called 'one below zero?'"
   - Lines 286 to 311, ending with Irene saying, "Nowhere."
   - Lines 312 to 325, ending with Carol's comment.

4. Consider your work on Question 3. What ideas about negative numbers are these students forming? Identify actions the teacher took or questions the teacher asked to promote this kind of thinking. Be specific and cite line numbers for each example.

5. What arguments do students in Case 19 use to support the idea that adding 3 and ⁻7 will produce ⁻4 regardless of their order? Are their arguments applicable to *all* pairs of integers? If so, explain. If not, how would you modify the arguments?

Fifth Homework

Reading assignment: Casebook Chapter 5

Read Chapter 5 in the Casebook, "Doing and Undoing, Staying the Same," including both the introductory text and Cases 20–25. As you read, note the generalizations that are present in students' work.

Write out the generalizations in words and/or symbols and bring them to the next session. These will be shared with other participants and also turned in to the seminar facilitator.

Writing assignment: Pursuing a mathematical question

This assignment is about the math *you* are learning in the seminar, not about the learning of your students.

Take some time before the next session to reflect on the math ideas from Session 4. What are those ideas? Which ideas make sense to you now? Which ideas are you still working on? How are you working on them?

Optional Problem Sheet 2

These problems will provide you with the opportunity to experiment with using symbolic notation. Feel free to turn in your work or questions on these problems at any time you would like comments or feedback. These problems will not be discussed during the regular seminar sessions. There will be an additional problem sheet available later in the seminar.

1. This is the same set of statements you considered in "Optional Problem Sheet 1." This time, assume that a, b, and n may be positive, negative, or zero and determine if the statements are always true. Which answers change and which do not? Explain each situation. Use story contexts, diagrams, or representations such as number lines and the charge model to support your thinking.

   a. $(a + b) - n = (a - n) + b$

   b. $(a + b) - n = (a) + (b - n)$

   c. $(a - b) - n = (a - n) - b$

   d. $(a - b) - n = (a) - (b - n)$

   e. $(a - b) - n = (a) - (b + n)$

   f. $(a + b) = (a + n) + (b - n)$

   g. $(a + b) = (a - n) + (b + n)$

   h. $(a - b) = (a + n) - (b - n)$

   i. $(a - b) = (a - n) - (b - n)$

   j. $(a - b) = (a - n) - (b + n)$

2. For each of the following equations, do NOT solve for x. Instead, reason out if x must be zero, a positive number, or a negative number. Use story contexts, diagrams, or representations such as number lines and the charge model to support your thinking.

   a. $x + 7 = 14$ b. $x + 7 = 7$ c. $x + 7 = 3$ d. $x + 7 = {}^-2$

   e. $x + 3 = 7$ f. $x - 3 = 7$ g. $3 + x = 7$ h. $3 - x = 7$

   i. $x + 3 = {}^-7$ j. $x - 3 = {}^-7$ k. $3 + x = {}^-7$ l. $3 - x = {}^-7$

   m. $({}^-x) + 3 = {}^-7$ n. $({}^-x) - 3 = 7$ o. $3 + ({}^-x) = {}^-7$ p. $3 - ({}^-x) = {}^-7$

REASONING ALGEBRAICALLY ABOUT OPERATIONS

Doing and Undoing, Staying the Same

Mathematical themes:
- What are ways to model and express the relationship between two quantities and their sum?
- How are the roles of 0 in addition and of 1 in multiplication the same? How are they different?
- How can we use the relationship between addition and subtraction to make sense of subtraction with negative numbers?

Session Agenda

Math activity and case discussion: Addition and Subtraction	Small groups	30 minutes
	Whole group	20 minutes
DVD: How addition and subtraction express the same relationship	Whole group	15 minutes
Break		15 minutes
Case discussion: Additive and multiplicative identities	Small groups	20 minutes
	Whole group	25 minutes
Math activity: Subtracting negative numbers	Small groups	30 minutes
	Whole group	20 minutes
Homework and exit cards	Whole group	5 minutes

Background Preparation

Read
- the Casebook, Chapter 5
- "Maxine's Journal" for Session 5
- the agenda for Session 5
- the Casebook, Chapter 8: Sections 6, 7, and 8

Work through
- Math activity and Focus Questions: Addition and subtraction (p. 180)
- Focus Questions: Additive and multiplicative identities (p. 181)
- Math activity: Subtracting negative numbers (p. 182)

Preview
- the DVD segments for Session 5

Post
- Criteria for Representation-Based Proof

Materials

Duplicate
- "Math activity and Focus Questions: Addition and subtraction" (p. 180)
- "Focus Questions: Additive and multiplicative identities" (p.181)
- "Math activity: Subtracting negative numbers" (p. 182)
- "Sixth Homework" (p. 183)

Obtain
- DVD player
- index cards
- cubes

Using Addition and Subtraction to Express the Relationship Between Two Quantities and Their Sum

Most elementary teachers are familiar with the language of "fact families" (or some similar phrase) to describe the set of equations implied by a single addition statement. Consider the example based on 3, 4, and 7: $3 + 4 = 7$, $4 + 3 = 7$, $7 - 4 = 3$, and $7 - 3 = 4$. These related equations are often called upon in elementary school mathematics as a tool to help students learn their addition and subtraction facts. In more general terms, these related statements can be written as "If $a + b = c$, then $b + a = c$, $c - a = b$, and $c - b = a$."

In RAO Session 5, participants examine models to explore how addition and subtraction represent different ways to express the relationship between two quantities and their sum. They first examine this notion with whole-number situations and then use that knowledge to help make sense of subtraction of negative numbers.

The first activity of the session involves examining four different problems using visual representations. Participants might model the four problems with cubes:

Peter caught 19 fish in the morning and 7 in the afternoon. How many fish did he catch?

$19 + 7 = 26$

Maeve had 7 red flowers and 19 yellow flowers. How many flowers does she have?

$7 + 19 = 26$

Elizabeth had 26 balloons, and 19 flew away. How many did she have left?

$26 - 19 = 7$

Codie had 26 cookies and ate 7 of them. How many did he have left?

$26 - 7 = 19$

Although the equation associated with each story situation is different, examining the four representations reveals that a single stack of cubes of two colors could model all of them. This stack provides an image of the relationship between 7, 19, and 26. It illustrates the statement, if $a + b = c$, then $b + a = c$, $c - b = a$, and $c - a = b$.

Later in the session, participants use the equivalence of $a + b = c$, $c - a = b$, and $c - b = a$ to examine subtraction with negative numbers. For instance, if we let $a = 5$, $b = {}^-3$, and $c = 2$ in the equations above, we obtain $5 + {}^-3 = 2$, $2 - 5 = {}^-3$ and $2 - {}^-3 = 5$. While participants might understand this is an implication of the relationship between addition and subtraction, they may still express doubt about the meaning of the equation $2 - {}^-3 = 5$.

Some participants might use the charge model to examine this problem. Point out that they should begin with a representation of 2 that includes several "pairs" of 0, i.e., several copies of $({}^+1 + {}^-1)$.

A representation of 2.   Now remove ⁻3.

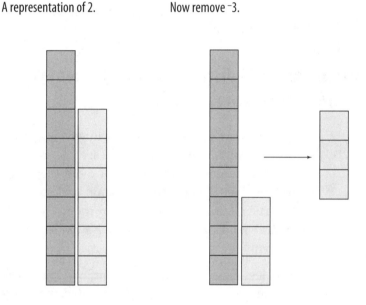

This leaves (⁺5).

Others might use a number line model by interpreting the subtraction expression $2 - {}^-3$ as showing the distance from ⁻3 to 2. In this situation they would place 2 and ⁻3 on the number line and determine that a move of 5 units to the right is required to move from ⁻3 to 2.

Still others may create story situations such as "The temperature was ⁻3 degrees at 6:00 a.m. and it was 2 degrees at 9:00 a.m. The change in temperature $2 - ({}^-3)$ is ⁺5 degrees."

For more information, "Maxine's Journal" (pp. 161–166) for this session includes a discussion of the ways participants explained why $5 - ({}^-3)$ is 8, using various representations.

Maxine's Journal

October 29

Parts of last night's session seemed very complicated. In a sense, there were two different ideas we were working on—inversely related operations and identities. But in fact, these two issues are closely connected, so I was working to show how they are tied together.

These two ideas seem, at first, like an interruption in the study of integers. However, we needed them—particularly the idea of how addition and subtraction are related—to get at the questions that arose in the previous session: What is the rule for subtracting negative numbers? How do you make sense of it?

Math activity and case discussion: Two different ways to think about the same relationship among quantities

In this first part of the session, I wanted the group to understand how addition and subtraction can both be seen to represent the same relationship among three quantities. The first problem on the Math activity and Focus Questions sheet "Examining addition and subtraction" is designed to make sure participants have that idea for themselves. Then they turn to the cases by Azar, Nadine, and Daisy to see how it shows up in first- and sixth-grade classrooms.

The last time I taught this seminar, participants had a difficult time understanding the relationship I wanted them to see among the four story problems—problems we might solve by calculating $19 + 7$, $7 + 19$, $26 - 19$, and $26 - 7$. Instead, they were looking at what different representations might highlight, such as which representations bring out the 10s and 1s more than others. So this time, I emphasized that I wanted them to read the four problems and think about the questions at the end of the second, third, and fourth problems that ask how the previous problem could help them. In addition, I wanted to make sure they used cubes and number lines to think about this. I added that these problems were given to a group of second graders, and so we are looking at these problems to see what second graders might learn about addition and subtraction.

Small groups

When participants started working on the problems, each group made a representation with cubes. Leeann, Madelyn, and Antonia arranged a different set of cubes for each problem. Each problem, however, was represented with a stack of 7 cubes and a stack of 19 cubes. Risa showed her group that you could make one stack of 26 cubes, 7 of one color, 19 of another, representing all four problems. Grace also wanted to see what the problems looked like on the number line. She said that she saw how the fish and the cookies problems would work together and then the flowers and the balloons problems. "Peter caught 19 fish in the morning," she said, pointing to 19 on the number line, "and 7 in the afternoon," showing a jump of 7 to land on 26. "Then, Codie started with 26 cookies," still with her finger at 26, "and ate 7 of them," moving back that same 7 to 19. "It reminds me of what students said in the introduction to the cases—you bounce up 7 and then you bounce back to land on the place you started."

When I got to Jorge, Risa, and James, they seemed mystified by Azar's case. What was so confusing to Patrice? Then James pointed out that maybe these students had never had an opportunity to think about the operations as students in Nadine's and Daisy's first-grade cases had.

Whole group

As I started the whole-group discussion, I decided to discuss the four story problems. I had a feeling that participants might learn something from the different representations. So I asked, "Did you have any new insights from building those representations?"

Denise said that she had understood the relationships among the problems when she first read them, but the members of her group had not. "It was only after Josephine and Vera built those cube stacks for each of the problems that I could actually *see* the idea I had in my head."

Madelyn showed her four arrangements for each of the problems. For the fish problem, she had a stack of 19 and a stack of 7 with a little space between the stacks; for the flowers problem, a stack of 7 and a stack of 19 with a little space in between. Madelyn said, "So looking at these two, we could see the connection. I kept seeing $a + b$ and $b + a$." I reminded participants of the term *Commutative Law* that we had discussed in the last two sessions.

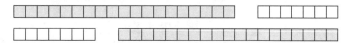

Madelyn went on, "For the balloons problem, we had 26 cubes with 19 separated out." She had a stack of 26 cubes lying down, with a gap between the 7 and the 19. "It looks almost the same as the flowers problem, doesn't it? And, when I look at the cubes for the cookies problem, they are just like the cubes for the fish problem."

Grace said she really liked the way all four problems were laid out by this group. "We used the same set of cubes for each problem. We could see that you never needed any additional cubes because the problems were using the same numbers, but because we had to take apart one problem to do the next, we didn't see the relationships among the problems the way you can see it with Madelyn's representation."

Denise said, "I think we really needed to have representations of the four problems to communicate in our group. I was trying to say how the fish and the cookies problems are related, but my partners didn't see what I meant."

Vera explained, "Denise was talking about how the fish problem and the cookies problem are opposites, but we didn't get what she meant. We thought she was saying that the fish problem and the flowers problem are opposites. When we worked with the cubes, we could see what she meant. It is like, in the fish problem, the numbers come together, and in the cookies problem, one of those numbers is taken away."

I asked, "So looking at the problems with the cubes helped all of you?" Everyone from that group nodded.

At this point, I wanted participants to see how just one representation could stand for all four problems. Because I had seen Yanique do this, I asked her to show the whole group.

Yanique showed us one stack of 26 cubes: 19 yellow, 7 red. "This can stand for any of the problems, it just depends on how you look at them. You can have the yellow cubes stand for the fish in the morning and the red stand for the fish in the afternoon. You can also have all 26 cubes stand for the cookies Codie had with the red cubes representing the ones he ate."

Vera and Leeann were able to explain to the whole group how Yanique's cubes show the flowers and the balloons problems. Then, I gave the whole group a moment to let the ideas sink in.

James said, "So, each of the problems is about the same number relationship. Actually, any one of those situations you can think of as addition or subtraction. How the numbers are related is the same."

M'Leah said, "The missing piece is just in a different place. The missing piece is either what you are starting with or what you are ending with. If you have the yellow cubes, and you are putting the red with it, you are going to have a larger amount. If you have all of the cubes and take the red ones away, you have only the yellow ones left."

Then Charlotte said, "If you take two parts and put them together, and then take one part away, you get the other part."

I asked, "So that is your way to generalize?"

Charlotte nodded and then Mishal spoke up. "If $a + b = c$, then $b + a = c$ and $c - a = b$, and $c - b = a$."

I wrote these four equations on the board and asked how participants were feeling about them.

M'Leah said, "In the past, I was always fine with algebraic notation, but now it has a whole different meaning for me. Now I can see in the notation the relationships we are talking about."

James said that he had written the relationship as $(a + b) - b = a$. "I didn't even have c."

June said, "When we get to algebraic notation, it feels like the ideas start floating away from me."

June was reminding me that many participants may need time to substitute numbers to fully understand—without that, the ideas would just "float away," to use June's expression. I suggested that everyone spend a few minutes to make sure they understood the notation. I asked that they first work on Mishal's four equations, and then James's.

DVD: How addition and subtraction express the same relationship

I told the group that we would watch a few minutes of a DVD segment of a second-grade class working with these same ideas. The class had been given the same set of four problems about fish, flowers, balloons, and cookies, and they had already talked about how the fish problem and the flowers problem are related. The DVD then depicted student discussions of the balloons problem and the cookies problem. I had not planned to spend a lot of time discussing the DVD segment, but I wanted participants to hear the students' language.

Before turning on the DVD, I reminded participants that we were watching it to think about students' ideas. There might be some things that are distracting in the DVD clip, but I hoped they would be able to set those things aside to think about the mathematical ideas at play.

After the DVD was finished, Jorge said, "Sawyer sounded just like Charlotte."

Leeann said, "It looked like a lot of students were not paying attention."

Lorraine responded, "I thought so, at first, too. I thought Sawyer was one of the students who seemed as if he weren't engaged, and then I realized he was the one who was talking! Maybe it's just the way students look, and you can't really tell."

Grace said, "I'm interested in the difference between the way Charlotte and Sawyer talk about the idea, and the way Mishal wrote it in algebraic notation. The way Charlotte and Sawyer say it, it sounds more like the way James wrote it: $(a + b) - b = a$."

James said, "We haven't said much about the print cases. At first, we didn't understand what was so confusing to Patrice. But, after our discussion—and after seeing this DVD segment and reading the cases written by the first-grade teachers—I see that there's something big to think about. If students go through school and never think about this, it's not surprising that it would be confusing."

Risa said, "It looks to me like only Sawyer and that girl understood it."

This is the kind of comment one often hears in reaction to watching the DVD, and I wonder what motivates it. Is Risa trying to dismiss the possibility that second graders can really take on this idea? Is it hard to watch a DVD of fidgeting students? Is Risa stating the question she would have if she were in the teacher's position? In any case, it is important to think about the evidence for making such a claim.

I responded, "From our position, watching this DVD clip and not being able to interact with students ourselves, we don't have an opportunity to get more evidence of what students do or do not understand. Karen, the teacher, has ended the lesson with something she wants her students to think about. So, this idea might not be finished in that classroom; she might return to it again and again. I would like you to think about what you would do if you were the teacher and you had Risa's concern. What would you do to find out about where the other students in the class stand with respect to this idea?"

We did not have time to pursue that question at this point, so I asked the whole group to think about it and called the break. We still had lots to accomplish in this session.

Case discussion: Additive and multiplicative identities

In the cases by Alice, Martha's question—looking at the equation $60 \times 1 = 60$, why does 1 give you back the same number? "I thought only 0 could do that!"—took students into a deep discussion about what the operations of addition and multiplication do. By examining what the operations actually *do* with the numbers that are operated on, they came to see why addition and multiplication require two different identity elements. Students were not using that language—"identity element"—but they had identified 0 as the additive identity (that is, $n + 0 = n$ for any value of n) and explored why 0 does not serve the same function for multiplication, but 1 does ($n \times 1 = n$ for any value of n).

These cases remind me of the earlier cases about switch-arounds. Although the teacher thought the question at hand was whether numbers could be exchanged in an addition problem and get the same answer, students seemed to be working with the question about whether the numbers can be exchanged in any problem. That is, they seem not to distinguish among the operations. Their sense is that "this is the way *numbers* work" rather than "this is the way *this particular operation* works." In Cases 23 and 24, students really seem to come to an understanding that it is not just a matter of how *numbers* work but of how the *operations* work.

Students in Alice's class come to understand that, because addition and multiplication behave differently, it makes sense that multiplication may need a *different* number to "do what 0 does for addition." They come to see that 1 does for multiplication what 0 does for addition. However, then they wonder, why do subtraction and division use the same numbers as addition and multiplication? We do not see if students ever get back to that question, but the answer can be found by applying the idea that the rest of the cases address. It is because addition and subtraction are so closely related in the way they act on numbers: $a + 0 = a$, and $a - 0 = a$. Similarly, though not addressed in the cases, multiplication and division are closely related, and both $a \times 1 = a$, and $a \div 1 = a$.

Anyway, those are the issues *I* see in the cases. Now, this is what happened in the seminar.

Small group

I first sat with Phuong, M'Leah, and Jorge. Initially, they were trying to figure out what was puzzling to students in these cases. After a while, Phuong said, "It does seem counterintuitive to end up with the same number you started with *after* you operate on it." That moved them into looking at what confused Martha and which comments in the classroom discussion seemed to help her.

Grace, June, and Yanique discussed Fran's ideas around line 406. Thinking about why, in multiplication, 1 gives back the same number, and in addition, 0 gives back the same number, she talked through what happens when 1 is *added* to a number and when 0 is *multiplied* by a number. They thought she was answering the question, Why are different numbers needed for the different operations to get back what you started with? To do that, she showed that when 1 and 0 are switched, the result is quite different.

Whole group

In the whole-group discussion, I first asked for the important idea students worked through in Alice's cases. Several ideas were voiced, which I wrote on the board:

- Numbers have characteristics for a particular operation
- Different operations treat numbers differently
- How numbers behave in different operations
- How 0 in addition has the same role as 1 in multiplication
- The parallel ways that operations work

These responses made me feel as if participants were getting to the heart of the two cases. First, I asked what James meant when he said numbers have characteristics for a particular operation. James responded, "Well, when you add 0 to any number, you get that number back. When you multiply 1 by any number, you get that number back."

I pointed out that these statements are as much about the operations as about the numbers. I also used this as an opportunity to share some vocabulary. "There are particular terms for these ideas: We say that 0 is the *identity element* for addition and 1 is the *identity element* for multiplication."

Some participants laughed a little at the language. The word *element* sounded odd to them. I also realized the word *identity* sounded odd, too, because it is used in such a different way in common speech—when someone thinks about an "ethnic identity," for example. In this example, *identity* is the name of a number with a special property.

I went on. "Two sessions ago, our questions about switch-arounds had us looking at, say, 4 and ⁻4, 7 and ⁻7, and we talked about the special language for that, too. We say that 4 and ⁻4 are additive inverses. We define the *additive inverse* of a number to be 'the number you add to the original number to get 0.' We say that ⁻4 is the additive inverse of 4 and 4 is the additive inverse of ⁻4 because ⁻4 + 4 = 0. Every number has an additive inverse. Also, every number except 0 has a multiplicative inverse. When you multiply a number by its multiplicative inverse, you get 1."

Risa reminded the group that the multiplicative inverse is also called *reciprocal*, which I acknowledged.

Going back to the list of ideas from the cases, I asked, "Tell me about operations treating numbers differently or numbers behaving differently with different operations."

Grace said, "Each operation does something different to the numbers. With addition, you bring two sets together. With multiplication, you have a certain number of groups that are the same size."

Grace's characterization of the operations is limited to whole numbers, but that is the domain that our group, as well as Alice's students, were discussing. Other participants continued in the same vein.

M'Leah said, "When you multiply two numbers, the numbers mean different things than when you add. This is what Karen was talking about around line 499. Like what she said, with 3×9. The 3 tells you how many groups of 9 you have. But when you add $3 + 9$, you just put 3 things and 9 things together."

Jorge added, "When you are adding, you have two amounts that are joined. When you multiply, it is as if that 3 is not an amount. It tells you how many times to take that amount of 9."

Kaneesha brought up another issue. She said, "I have to say, I got really stuck on line 325: $1 \times 60 = 60 + 0$. Martha said that helped her a lot, but I don't see it."

I was not sure what Kaneesha was having a problem with, but several participants indicated they were also mulling over that line. Was it that they had a problem interpreting the equation for themselves, or was it that they did not see how it could give Martha insight?

What seemed to confuse participants is the idea of using multiplication and addition on different sides of the equal sign. The discussion continued for a few minutes until Madelyn suggested that they might look at the equation as $1 \times 60 = 60 + (0 \times 60)$. At that, Kaneesha exclaimed, "Yes! That's it!"

Charlotte said, "When I look at it written that way, as Madelyn did, it seems to be a statement of what 1×60 means, which is what Martha had been confused about in the case. It says that when you multiply 1 by 60, you have one 60, and you don't add any more 60s, or you add zero 60s."

Although the group now seemed satisfied with their conclusion, I was still bothered that they could not simply accept $1 \times 60 = 60 + 0$. I went back to that equation and said, "The equal sign simply means that the expression on either side of it has the same value. We could also say $2 \times 60 = 80 + 40$, or $1 \times 60 = 38 + 22$."

Now I brought up one last idea. "I want to look at line 514, at the end of Alice's second case. Mary says, 'If multiplication and addition need different numbers to do that, why do subtraction and division just use the same ones that addition and multiplication use? Why don't they all have different numbers? How come they just use 0 and 1 again?' Do you have any thoughts about that?"

M'Leah said, "I bet it has to do with what we were talking about before—about how addition and subtraction undo each other."

I suggested we look back at those algebraic equations Mishal had given us. "Let's look at $a + b = c$ and $c - b = a$. Before you spent some time checking out what happens when you substitute different numbers in. Now, work with your small groups to see what happens if $b = 0$. "

June immediately called me over. She had written "$a + 0 = c$" and "$c - 0 = a$." She asked, "What does that give you?"

Denise immediately spoke up. "If you know that b is zero, then you know that $a = c$. So then it becomes $a + 0 = a$ and $a - 0 = a$. Neat."

June and Vera looked quite uneasy. They were not going to understand the idea from the algebraic notation. So, I suggested that they go back to Case 16 from Chapter 4 in which students talked about how they represented $12 + 0$ with cubes. They went back to review that case, but I brought the whole group back together before they finished thinking through the idea.

I said that some participants were able to think through the idea with algebraic notation, and I presented the argument to the group in the same way that Denise had explained it in her small group. I then acknowledged that not everyone connected with the algebraic notation. In order to think it through, we should refer to a different representation. As I had said to June and Vera, we might think about Case 16 from the previous chapter, in which the first graders discussed what $12 + 0$ looks like with cubes.

305

M'Leah said, "Oh, yeah. Cool." When I asked her to explain what she saw, she said, "The first graders had this idea of a place at the end of the tower that held 0 cubes. So we can think of it sort of like Yanique's model before—in which the one stack of cubes can be seen as representing both the addition problem and the subtraction problem, depending on how you look at it. So now, you have a stack of some cubes, and a place at the end that holds 0. Let's say there are n cubes. So it can show $n + 0 = n$, if you are thinking about those 0 cubes as being added on. But, if you start out thinking of the n cubes and 0 cubes being there from the beginning, then you take away those 0 cubes and you have $n - 0 = n$."

I looked over at June, who was smiling broadly, and Vera, who was smiling a little more tentatively. Then Kaneesha said, "I want to use our charge model. This way, I can represent zero as 1 red cube and 1 blue cube. No matter how many cubes I start with, if I add the red and blue cubes, it still stands for that amount. When I take the red and blue cubes away—I still have that same amount." Now Vera was smiling more assuredly.

It seemed like everyone was feeling pretty solid with this idea now. "We don't know if Alice's students ever got back to their question, but it seems like we have a good idea about it. We have been talking about how addition and subtraction actually look at the same relationship among three numbers—but from different perspectives. Both operations are about how two parts make up a whole. We have been looking at this special case of when one of those parts is 0. We might say, the other part is equal to the whole. Depending on how you look at it, $n + 0 = n$ and $n - 0 = n$."

I still needed to clarify one point: "There's also something important about the difference between addition and subtraction that is relevant here. Obviously, we know that $n + 0 = n$ and $0 + n = n$. By definition, for a number to be an identity element, it needs to work in both situations. But we know that unless $n = 0$, $0 - n = {}^-n$, and not n. So, in that way, 0 does not behave exactly the same in addition and in subtraction. In subtraction, 0 is only a one-sided identity."

Everyone was nodding, so I went on. "Let's go back to $a + b = c$ and $c - b = a$. The next question for us to consider is, What if one of those numbers is less than 0? Let's look at $8 + {}^-3 = 5$. So $a = 8$, $b = {}^-3$, and $c = 5$. I have written out an addition equation. What are the corresponding subtraction equations? How do they help us understand subtraction of negative numbers?"

This was the math activity we would be working on for the remainder of the session.

Math activity: Subtracting negative numbers

Integer arithmetic is difficult for lots of people because it is the first topic in the study of the number system that is not motivated by obvious, natural models. Indeed, it took thousands of years for negative numbers to be accepted as part of the number system. Ancient Chinese people worked with red and black rods but would always restate answers using a positive number, a practice we continue today. (Have you ever really said, "I have $⁻4," or do you rephrase it as "I owe $4"?) Temperature and elevation are among the very few situations in which we actually see negative numbers; but even there, we usually say "5 below zero" rather than "⁻5 degrees," or "128 feet below sea level" rather than "⁻128 feet above sea level."

The "rules" for integer arithmetic are motivated by mathematical consistency: Through centuries of experience, it became apparent that the Distributive Rule for Multiplication over Addition is central to our number system, and so $0 = ⁻2 \times 0 = ⁻2(3 + ⁻3) = ⁻2 \times 3 + ⁻2 \times (⁻3)$. In other words, $⁻2 \times ⁻3$ must be the opposite of $⁻2 \times 3$. Similarly, because subtracting a whole number "undoes" adding the same number (e.g., $6 + 3 - 3 = 6$), integer arithmetic is defined in such a way as to make the same fact true in the new context: $(8 + ⁻3) - (⁻3) = 8$, or $5 - (⁻3) = 8$; similarly, $⁻3 - 5 = (⁻8 + 5) - 5 = ⁻8$.

We try to build models for integer arithmetic to have some representation that is less abstract, but these models are not as obvious as the whole-number models. They were not as obvious to my participants last night either.

Small groups

Everyone moved into the work with lots of energy. Within a few minutes, they all had the two equations that correspond to $8 + ⁻3 = 5$, $5 - 8 = ⁻3$, and $5 - (⁻3) = 8$. Antonia said, "It is clear that $5 - 8 = ⁻3$. We have been talking about that for weeks. But, $5 - (⁻3) = 8$. That is kind of mind-blowing."

Kaneesha and Vera relished the opportunity to think more with the charge model from the last session, and now working in different groups, showed their partners what they had seen in Session 4. If blue cubes represent the positives and red cubes represent the negatives, and you start with 5 blue cubes, how do you take away 3 red cubes? To do that, you first have to add 3 blue cubes and 3 red cubes, which make 0. So, once you add on that form of 0, you can take away 3 red cubes, and you are left with 8 blue cubes. I went by as Kaneesha finished her demonstration. She looked up and giggled. "I love it," she said. (See p. 151 for an illustration of a similar problem.)

The rest of the group was not giggling but was intrigued. They all picked up cubes and began representing different subtraction problems.

Some groups stuck with this model for the duration of the small-group work, and I decided not to push them. Clearly, this was helping them think about what happens when you subtract negative numbers. A few groups went on to explore the number line.

Antonia was showing her group that they can look at subtraction as the distance between two points. "If you look at the distance between 5 and $^-3$, you get 8. So, $5 - (^-3) = 8$."

James pointed out that you might think of a context. "If I start the day with $5 and at the end of the day owe my mother $3, how much money did I spend?"

Madelyn said, "OK, I see that. But now, how do you look at subtraction as distance between two points and get $^-3 - 5 = ^-8$?"

That question that Madelyn was asking was a good one. There is something tricky, when the system expands to include negative numbers, about using a definition of subtraction that implies positive numbers. In fact, distance is the *absolute value* of the result of subtraction. However, if the result is already positive, you do not have to worry about that. Now that we have expanded the subtraction problems under consideration, in order to think about distance as subtraction on the number line, you have to include direction, too. If you look at $5 - (^-3)$, think about what you need to add to $^-3$ to get 5. If you look at $^-3 - 5$, think about what you add to 5 to get $^-3$. In that sense, we are right back to the relationship between addition and subtraction. Here we are thinking of the subtraction as finding a missing addend.

M'Leah, Grace, and Risa called me over to talk about the stories they had made up. Grace said, "I have 8 students. Today there are 5 present, and so I mark 3 absent. A few minutes after I do this, a late bus arrives, and the 3 absent students enter the classroom. I remove the absences—I subtract $^-3$—and so now I have 8 students present."

Risa said she was not really sure if that matched the number sentence. Grace said she thought it did. "You have 5 students in the room and subtract the 3 absences: $5 - (^-3) = 8$." Risa said, "I don't know. I just don't see that $^-3$ as being in the room."

I asked Risa if she had a story. She said yes, but it does not seem to work out. "I have a $5 bill and three $1 bills. I owe my friend $3. So my net worth is $5. Now my friend cancels my debt so I keep the $3. My net worth is $8. The only problem is that you had to know the answer was $8 to set it up right. If I say I have $5 and my friend forgave me a debt of $3, this feels more

Doing and Undoing, Staying the Same

like 5 − (⁻3) is $5. I had $5 to begin with and owed $3; now I do not owe $3, so I am left with $5."

I said, "OK. Let's think about this situation. You have $5, but you owe $3. So what is your net worth? If you think about your financial status right at that point, how much do you really have?"

Now, Risa seemed to get it. "Ohhhhh. At that point, my net worth is $2. So, when the debt is forgiven, I go up to $5. So that's the story for 2 − (⁻3) is 5. Right."

M'Leah said she still was not sure, but she would think about it. I left the group so that the three could continue talking about the stories.

Then June called me over and told me that she was completely befuddled. Yes, she understood how to use the charge model to think about subtracting negatives, but the number line was not making sense at all. "I am trying to let go of previous assumptions I have had about negative numbers, but I am not sure which ones to hold onto and which ones to let go of. To me, the minus sign has always meant 'go backward.' So 5 − (⁻3) means you start at 5, and with those two minus signs, *make sure* you go backward 3. That lands me on 2. How on earth do you get to 8?"

June's thinking about subtraction on the number line was not in terms of finding the difference. To her, 5 − 3 is represented as take a jump of 3 back from 5.

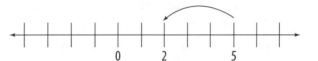

I acknowledged that it is kind of crazy that subtraction can land you farther ahead from where you start. Then I told her that I have a way of thinking about this with the number line that some people like a lot and some people hate. "So, here it is. You have to think as though you are standing on the number line. If you are doing 5 − 3, think of yourself as standing at 5 looking forward. Subtraction means to turn around and step forward—remember you are now facing toward the smaller numbers—3 steps. You land on 2. You've gone backward on the number line.

"And when you add a negative number, that also means you go backward on the number line. For 5 + (⁻3), think of yourself as standing at the 5 looking forward. Addition means you keep looking forward, toward the larger numbers, but the ⁻3 means you step backward 3 steps and land on 2.

"Now consider subtraction of a negative number: 5 − (⁻3)." You are standing at 5 facing forward. Subtraction means to turn around, so you are facing the smaller numbers. But now you are subtracting ⁻3, so that means after you've turned around to face the smaller numbers, you step backward 3 steps. That lands you on 8!

"What has to happen is that you need to stretch your interpretation of the moves on the number line. If you can separate what it means to subtract a positive and what it means to add a negative, then you can combine those meanings to see what happens when you subtract a negative."

June said that she kind of liked that, but she needed to talk it through with other numbers. I left as she started to explain to Vera how to work on $10 - 1$, $10 + {}^-1$, and $10 - ({}^-1)$ on the number line. When I checked with her and her partners a few minutes later, they said they were doing fine. Now they were trying ${}^-10 - ({}^-4)$.

Yanique, Vera, and Lorraine were working to come up with a new context for subtracting negatives—gaining and losing yardage in a football game. "Adding and subtracting has to do with time," Yanique explained to me. "Let's say your team is at its 30-yard line, and in the last play, they lost 5 yards. If you ask where they were before the last play, you take away that ${}^-5$: $30 - ({}^-5) = 35$." As I listened, it became clear that this context could work well. The group had separated what positive and negative numbers stand for (gaining and losing yardage) and what addition and subtraction are (finding where the team is after the next play or was before the last play). However, as I started asking them questions, I realized that the process was still somewhat muddled for them; so I presented four contexts for them, for which they wrote the corresponding arithmetic sentences.

- The team is at its 30-yard line. In the next play, they gain 5 yards. Where are they after the next play? $30 + 5 = 35$
- The team is at its 30-yard line. In the next play, they lose 5 yards. Where are they after the next play? $30 + {}^-5 = 25$
- The team is at its 30-yard line. In the last play, they gained 5 yards. Where were they before the last play? $30 - 5 = 25$
- The team is at its 30-yard line. In the last play, they lost 5 yards. Where were they before the last play? $30 - ({}^-5) = 35$

With that, they felt pretty satisfied. What is problematic with this context, though, is that the football field is not like a number line. If you start at one goal, the yards stretch from 0 to 50, and then they back down to 0. In order for this representation to work, you have to think about only half the field. It works if the team with the ball is on its 30-yard line. If they are on their opponent's 30-yard line, the representation does not work as a model for adding and subtracting integers.

Whole group

When I brought everyone together for a whole-group discussion, I asked that they talk about what they were learning by using the number line and the charge model.

Most participants said they felt the charge model works better than the number line for thinking about subtracting negatives. Though M'Leah said

Doing and Undoing, Staying the Same

she still had a problem with the charge model. "It feels like you have to already know the answer to solve it."

Then Grace raised her hand. "I know what M'Leah is saying, because that was bothering me for a while, too. But then, here's what I was thinking." She held up two large towers of cubes.

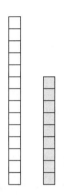

"This shows how much money you have in your account and how much money you owe. You don't necessarily know how many cubes are in the two towers; you just know that the 'have' tower (light blue) is 5 more than the 'owe' tower (red). That means that you have $5 available to spend. So, your net worth is $5. Then one of my creditors comes to me and says, 'You know those $3 you owe me? You can forget it; you don't have to pay me back.' So a $3 debt is forgiven, and you can take it away."

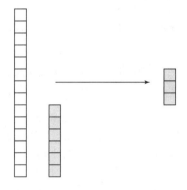

Grace finished, "And now you see you have $8 available to spend."

M'Leah sat watching carefully. Then she nodded. "Yes, that image makes a lot more sense to me."

Then James said, "I want to see something else. Grace, put back those three red cubes. You're back to the original situation; you have a net worth of $5. Now, let's say someone gives you $3, so you add 3 blue cubes to your 'have' tower."

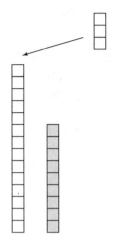

Grace added the 3 cubes, again showing a net worth of $8.

James explained, "It is not the same, but it gives you the same net worth, subtracting ⁻3 or adding 3."

M'Leah said, "OK. I am satisfied. That model works for me."

Mishal pointed out that when they were talking about which was larger—⁻9 or 7—the number line worked much better. The charge model did not help when comparing integers; it was a lot easier just to look at the number line and think about which number was to the right.

I nodded and said, "It's important to understand that any representation highlights some ideas and obscures others. That's why it is so important to work with a variety of representations."

June said, "I know what you're saying, but at first, I was completely befuddled by subtracting negatives on the number line. Luckily, Maxine showed me how to think about it, and now I really like it."

Now others wanted to hear more from June, but we were out of time. Maybe my decision to hold off on the whole-group discussion had been a mistake. I suggested that, after responding to exit cards, they might spend a few minutes talking to June, or they could send June or me an e-mail message, and we could continue to work on this idea electronically.

Exit cards

In Session 1, I was aware that many participants were intimidated when they saw algebraic notation, and others were quite comfortable with it. So now, more than halfway through the seminar, I asked participants to write on their exit cards how they were feeling about that notation. I also asked a second question about what they were getting from my written responses to their homework. The following are answers to the first question:

YANIQUE: "Growing up, I had no problem 'plugging in' numbers to the variables. It became sort of a game to figure out what the variables stood for. I have realized that it made no sense to me then—I was just solving a puzzle. Now, thinking about the concepts makes more sense, and I can substitute any variables for the generalizations I am making. I feel that I am working backward, numbers first, then the variables."

545

LORRAINE: "I find that the algebraic notations are helpful in seeing the generalizations that we have made. I like seeing the 'rules' that we have come up with in class written in such a way that they can be readily understood."

550

CLAUDETTE: "As long as I can generate the formula myself and use different numbers to see that it is actually working, I am OK with it. I feel like I have some ownership over the formula."

555

JAMES: "I enjoy the algebraic notation. I feel it makes it easier to deal with everything when it is all said and done. Also, you usually come to these generalizations after you work through something and come to a conclusion, which means that when you are developing a notation, you are doing so out of an understanding of a mathematical situation. I think that's great."

560

DENISE: "Actually, I like the use of algebraic notation. Sometimes, in a way, it makes the math or the relationship clearer when specific numbers are not used. It is like the relationship stands on its own or is apparent without the use of specifics. But then, in the same way that different representations work more or less effectively in different situations, it can sometimes be helpful to go back and plug in some numbers to 'see' what is happening."

565

570

PHUONG: "I noticed that it is helpful if it has been presented visually.

575

"Just noticing the part-whole relationship helps me generate a whole bunch of equations":

$$A + e = N$$
$$N - e = A$$
$$N - A = e$$

580

JORGE: "I like trying to use a model to explain the notation. Instead of just following steps, creating situations helps me to understand concepts."

M'LEAH: "I understand the algebraic equations when dealing with 'easy' numbers. Yet, I become very fuzzy when dealing with negative numbers in algebraic equations. What does $^-a - {}^-b$ really equal? Is it ^-c or just c? I am *so* confused! P.S. Do not worry about the confusion. I will work it out over the next few weeks. My confusion is in a good way."

CHARLOTTE: "It only feels OK when I am translating math work I have just determined. So it works if I am in charge of assigning the variables to my work. It still feels bad (i.e., brings up issues of insecurity, sadness, fear) when I have to *interpret variables*. I cannot rely on theory—or experience—I still need to translate that notation into something I *can* interpret."

Answers to the second question, about what participants get from my written responses to their homework, were very reassuring.

PHUONG: "Your feedback helps me to be more thorough and attentive when I respond to the student-thinking assignments."

JAMES: "As for the correspondence, I love it. It is much more than I expected, and I am extremely grateful for the ideas it poses and the questions I can derive from it and see as my own ideas. I also get ideas I can bring into my class."

JORGE: "Your feedback gives me an objective look at what I am doing and other considerations for my reflections. Thank you."

M'LEAH: "The responses have been extremely helpful revolving around the activities I do with my students. There are many more mathematical ideas that I can discuss with my students. The responses are helping me search for those ideas."

YANIQUE: "I enjoy reading the feedback for several reasons: The ideas give me more things to think about, new approaches, etc. Your feedback also rephrases what I am thinking and helps me understand a bit more."

JOSEPHINE: "I find it quite supportive and motivating because someone takes time to make sense out of what I

Doing and Undoing, Staying the Same

do not understand and clarify it. I really appreciate your work. I feel that I will do the same for my students." 625

KANEESHA: "I find that it offers more food for thought, provides insight into whether I am on the right track, and allows for more of a connection with the facilitator. It is nice to know that a struggle is understood and can be directed into clearer thinking on behalf of the participant." 630

RISA: "I find the responses helpful. I know that I tend to want all students to understand, but I often don't consider that not everyone has the same level of understanding. So, your comments and suggestions make me slow down and allow time for students to think through the ideas. I also find it helpful and encouraging that you see a positive change in my questions. I know that I am more aware of my questions because of your guidance. I am now asking questions that allow my students to explore more. Thank you." 635 640

CLAUDETTE: "I especially benefit from suggestions and questions. I have to think a little deeper about my mathematical thinking as well as my students do." 645

Responding to the fifth homework

November 1

For homework, the participants wrote about the mathematics of Session 4, comparing integers.

Mishal 650

Last week, we started out explaining how we would show that 9 is greater than 7. As a group, we came up with many of the same ways as was listed on the board—counting, adding on, number-line position, distance away from zero, one-to-one match, tower comparisons. When we moved into negative integers, it became more difficult to make the same kinds of comparisons. 655

Comparing negative integers with positive integers was not a problem with a number line. I found that it was more difficult seeing the comparison when towers were used because we had to create the idea that positive values would be one color and the negatives values, another. When these colors were understood, it was easy to see the comparison, but I felt that because these colors are arbitrary and changeable that this method was not dependable. 660

Comparing negative integers with negative integers presented more of a prob-
lem. I tried the number line, charges, and towers. Again, the number line
seemed to be the most reliable to show comparison, but then another problem
arose. When comparing ⁻5 and ⁻3, numerically 5 is larger than 3, but when it is
a negative integer, 5 is smaller than 3. This concept seems to shake my sense of
numbers. I was glad that someone mentioned it in terms of temperature:
If the temperature is ⁻3 degrees, it will be warmer than ⁻5 degrees. The concept
of owing money was also brought up. I knew that if I had to pay off a debt
of $3, I was better off than when I had to pay off a debt of $5. Using the
negative charges and matching them up would not indicate which number
had the greater value.

I think that when negative integers are used, one needs to think a bit differently.
Moving to the left of the zero on the number line causes the value of the numbers
to be the reverse of the numbers to the right. (⁻1 is bigger than ⁻2 and ⁻3, but 1 is
smaller than 2 and 3.) I feel that, for me, the positive integers are more standard,
familiar, and safe than negative integers.

I wonder if it is possible to use a concrete model similar to the towers to show
comparison between negative integers.

In this writing, Mishal is working on how to make sense of comparing negative numbers that, in some ways, feels contradictory to the way whole numbers are compared. In particular, she needs to think about why ⁻5 is less than
⁻3 when 5 is greater than 3. (Mishal's words—"When comparing ⁻5 and ⁻3,
numerically 5 is larger than 3, but when it is a negative integer, 5 is smaller
than 3"—are pretty confusing, but I think that is what she means.) She uses
various contexts and models to think about it and expresses her unease. In
my response to her, I wanted to reassure her that this discomfort is quite
natural and has to do with the strangeness of new ideas but that her thinking
is strong. If she continues to think about negative numbers, she will get used
to them.

Dear Mishal,

Absolutely, positive numbers are easier to think about than negative numbers.
We first learn about numbers through counting objects, and there are certain
ways those numbers behave that we can rely on. The thing is, as we go through
school (and this happened historically, too), we are introduced to different kinds
of numbers, and when we do that, we have to rethink what we understand.
The first big jump of that kind for students is coming to think of 0 as a number.
How can it be a number when there is not anything to count? In order to
consider 0 a number, students need to expand their sense of what a number is.

Something similar happens when students start working with fractions.
If you have a cookie broken into 3 equal parts and you eat 2 of them, you
can count the 1 cookie, you can count the 3 parts in the cookie, and you can
count the 2 parts you ate. But then there is this new kind of number that has
to do with holding onto the idea of the whole cookie and putting those parts

Doing and Undoing, Staying the Same

together: $\frac{2}{3}$. When students are first introduced to fractions, some of them 705
find that mind-blowing. I remember working with one third grader who had
the hardest time putting those ideas together and holding onto them to think
about what the quantity is, much less operate with those new numbers.

Anyway, the same thing happens with negative numbers. Suddenly we need
new ways to think about what these numbers mean and how they operate. In 710
the same way that we got used to thinking about 0 and fractions as numbers,
if we work with negative numbers enough, we will get used to them, too. At
first, it is bound to make us feel a little shaky. Still, you should feel assured
that what you have written is correct. You should also know that when you
are working with negative numbers, you can take them back to contexts and 715
models that help you think. When confronted with a "naked number" problem
with integers, you can see if the numbers could fit a word problem about
temperature or a word problem about having and owing money. You might
see if the problem can be represented on a number line. If it helps, you can
also think about whether the problem could be represented with the model 720
about charges (even though you pointed out some of the limitations of that
model).

Mishal, I cannot quite tell if you are aware of what a strong mathematical
thinker you are. You often write about feeling shaky or being thrown off
balance, but your ideas are right on target. 725

Maxine

Detailed Agenda

Note: There are three ideas to explore in this session:

- The relationship between addition and subtraction
- The roles that 0 and 1 play in the number system
- Making sense of subtraction when numbers are negative

While the agenda is divided so that approximately 1 hour is spent on each idea, the first two ideas may require additional time. If that is the case, set aside the last 30 minutes to allow participants to work on the "Math activity" sheet, "Subtraction with negative numbers," in small groups and omit the whole-group discussion of this topic.

Math activity and case discussion: Two different ways to think about the same relationship among quantities   (50 minutes)

Small groups (30 minutes)

Whole group (20 minutes)

Distribute the Session 5 Math activity, "Addition and subtraction." Call attention to the directions for Problem 1 by explaining that acting out the story with cubes is a way to highlight the actions involved in each story situation. Let the group know that they will examine the math questions and discuss the first three cases in this period. You can also remind them to share the generalizations they wrote about these first three cases when they examine the Focus Questions.

As you circulate among the small groups, take note of how they represent the four problems to determine what models should be shared during the whole-group discussion. Be sure that a number line and a cube model are among those discussed.

As participants are building models, ask how the actions with the cubes or the number lines match the story context. If participants notice that a single stack of cubes can model all four problems, ask them to explain how they interpret the model for each problem. If no participant brings this up, ask the small group to consider how each model is similar and how it is different from the other. Encourage them to consider what generalization(s) their models support (for example, if $a + b = c$, then $c - b = a$, or if $a + b = c$, then $b + a = c$).

After 15 or 20 minutes, remind the participants that they should also be looking at the first three cases. These cases provide examples of students who are examining how both addition and subtraction can represent the same situation. As participants work on these questions, ask how the thinking of students is related to their own models of the four problems.

As participants work on the Math activity, they should talk through the arguments offered by students in Case 20 and identify connections between the ways students in Cases 21 and 22 solve the problems and the generalizations the participants made in response to the math activity. If there is time, participants can also discuss how the ideas in these Grade 1 cases are revisited by students in sixth grade (students in Case 20).

Begin the whole-group discussion by soliciting models of the fish, flowers, balloons, and cookies problems. Participants should share a variety of models including number-line and cube models. As the models are offered, ask questions so that participants are connecting the actions of the story context with the representations.

If some participants noted that a single model could be used and reinterpreted for each of the four problems, ask them to illustrate their thinking. If no participants recognized this, ask how the models for pairs of problems, such as fish and balloons or cookies and flowers, are alike and how they are different.

Some participants might use the term *fact family* to express the general relationships. For instance, $3 + 4 = 7$, $4 + 3 = 7$, $7 - 4 = 3$, and $7 - 3 = 4$ form a fact family expressing the relationships among 3, 4, and 7. If the term *fact family* comes up, ask participants how that term is related to the problems using 26, 19, and 7.

Once participants are comfortable using words to state their generalizations connecting addition and subtraction, introduce the symbolic notation, if $a + b = c$, then $c - b = a$ and $c - a = b$. Once this notation is introduced (whether by a participant or by you), give participants a few minutes to try various numbers in place of a, b, and c so that they may become familiar with what the notation is expressing. (*Note*: Because this notation will be used later in the session to examine subtraction of negative numbers, it is important to establish the meaning of this statement now with positive numbers.)

If you have time, address Question 3 of the Math activity by asking, "What opportunities for developing ideas about the relationship between addition and subtraction do students' comments provide, and how can teachers take advantage of such opportunities to develop these ideas further?"

DVD: How addition and subtraction express the same relationship (15 minutes)

Whole group

Introduce the DVD of a group of second graders discussing the fish, flowers, balloons, and cookies problems. Ask participants to listen carefully and take notes about how the students are thinking. (When this DVD segment was recorded, there was a problem with the microphone. Participants will have to listen through the scratchiness.)

Play the DVD segment once and then ask for comments about Sawyer's statement, "If two numbers come together and one goes away, then the one that is left is the answer." To begin the discussion, ask questions such as "What generalizations do you hear in this student's statements?" "What evidence do you see in students' statements that they are noting a general relationship and not just a single example?" "What is the connection between Sawyer's comment and the work we are doing examining the relationship between addition and subtraction?"

Representing a Student's Response

In the DVD clip, Sawyer says, "When two numbers come together and one goes away, the one that is left is the answer." There are different ways to represent Sawyer's statement using symbols. One possibility is if $a + b = c$, then $c - b = a$ or $c - a = b$. Another way to represent his words might be $(a + b) - b = a$ and $(a + b) - a = b$.

The differences between these two versions highlight an issue that was raised in RAO, Session 2. In the first, c is used to express the sum of $(a + b)$. In the second, the single expression $(a + b)$ represents both the instruction, "add a to b," and the result of that addition, the sum of $(a + b)$.

Break
(15 minutes)

Case discussion: Additive and multiplicative identities
(45 minutes)

Small groups (20 minutes)

Whole group (25 minutes)

Cases 23 and 24 explore the differences and similarities between addition and multiplication through examining the role that 0 and 1 play in each operation. Examining students' questions about what 0 "does" and what 1 "does" allows participants to consider the general structures of addition and multiplication.

Distribute the "Additive and multiplicative identities" sheet. Ask, "What are the roles of 0 and 1 in the number system?" Let participants know they will have 20 minutes to discuss these questions with their group.

Understanding Identity Elements

In the RAO seminar, participants encounter general mathematical principles that have been present but not highlighted in their understanding of arithmetic. The two statements "1 × any number = that number" ($1 \times n = n$) and "0 + any number = that number" ($0 + n = n$) are good examples. Neither of these statements will be surprising or unfamiliar, but what may be new to participants is the idea that each of these represents a more general relationship. Stated in more formal mathematical terms: For a given operation (which can be symbolized as Θ), if there is a number, a, so that $a \Theta n = n$ and $n \Theta a = n$ for any value of n, then a is the identity element for that operation. Considering the two operations of addition and multiplication, we can see what is the same (they both have an identity element) and what is different (1 is the identity element for multiplication and 0 is the identity element for addition).

In the case of subtraction and division, 0 and 1 play a similar but more limited role. $n - 0 = n$ and $n \div 1 = n$; however, reversing the order of the terms changes the result. In other words, for most values of n, $0 - n \neq n$ and $1 \div n \neq n$. Therefore the operations of subtraction and division do not have identity elements.

Question 1 provides an opportunity for participants to examine and articulate the main point of this discussion, that 0 in addition "acts like" 1 in multiplication. While the formal mathematical terms for these relationships—that *0* is the *identity element* for addition and *1* is the *identity element* for multiplication—will be a part of the whole-group discussion, you might also want to provide this language to some small groups.

In regard to Question 2, some participants might need to discuss the meaning of the equal sign in the statement "$1 \times 60 = 0 + 60$." They might be more familiar with equations in which it is possible to see how one expression can be transformed into the other, as in $1 + 6 = 2 + 5$. This question is a way to initiate a discussion about the meaning of the equal sign. It is simply a notation to indicate both expressions have the same value. $1 \times 60 = 0 + 60$ is true because both expressions have a value of 60.

Question 3 extends the discussion by including the operations of subtraction and division. Participants will explore the ideas that "any number minus 0 is that number" and that "any number divided by 1 is that number."

Begin the whole-group discussion by asking participants to list the important mathematical ideas that students were working on in the two cases written by Alice. An important principle is that the focus is on the operations themselves as entities to be studied and not about the results of computation or purely numerical work. See "Maxine's Journal" (p. 157) for an example of a possible list and the discussion it provoked.

Use the mathematical ideas from this list both to clarify how addition and multiplication are different and to note their parallel structures ($1 \times n = n$ and $0 + n = n$). Introduce the term *identity element*. You might also want to revisit the conclusions from Session 3, [$(a - b) + (b - a) = 0$ and $\frac{a}{b} \times \frac{b}{a} = 1$], as a further example of the parallel structure and a further explication of the terms *identity* and *inverse*.

Near the end of the whole-group discussion, turn to Question 3 and the question Mary raises in the case, "If multiplication and addition need different numbers to do that, why do subtraction and division just use the same ones that addition and multiplication use? Why don't they all have different numbers? How come they just use 0 and 1 again?"

As participants share their ideas, look for ways to connect their ideas to what was explored in the first part of the session—that addition and subtraction both express the relationship among two quantities and their sum. You might suggest they look at the statement "If $a + b = c$, then $c - b = a$" and consider what happens if $b = 0$. (See the discussion in "Maxine's Journal" pp. 159–160.) Participants can explore this idea with numbers, cubes, story contexts, and symbols.

Once this idea is discussed, tell the group that in the next part of the seminar they will revisit the statement, "If $a + b = c$, then $c - b = a$" to examine what happens when b is less than 0.

Math activity: Subtracting negative numbers   (50 minutes)

Small groups (30 minutes)

Whole group (20 minutes)

Begin the Math activity by working through Problem 1 as a whole group. First point to the statements that were written as a result of the earlier work on addition and subtraction: If $a + b = c$, then $c - b = a$ and $c - a = b$. Next, consider $a = 9$, $b = 4$, and $c = 13$. Have the group use these values to write out the three statements: $9 + 4 = 13, 13 - 4 = 9$, and $13 - 9 = 4$. Then ask the group to let $a = 8$ and $b = {}^-3$ and consider what the corresponding statements would be. Give participants a few minutes to work with someone sitting next to them; then solicit the statements: $8 + {}^-3 = 5, 5 - {}^-3 = 8, 5 - 8 = {}^-3$. Explain that this method of relying on mathematical consistency is one way to think about subtraction of negative numbers.

Distribute the "Subtracting negative numbers" sheet and invite participants to continue to explore subtraction of negative numbers by drawing on story contexts and the two models of negative numbers they have been using: the charge model and the number line.

Exploring Types of Subtraction

In the DMI module "Making Meaning for Operations," participants explored the connections between various story situations and subtraction. They saw that story contexts for subtraction include not only take-away situations but also situations involving comparison and missing addends, among others. When working to make meaning for subtracting negative numbers, it is helpful to consider a variety of subtraction situations.

For instance, $5 - 3$ could be an arithmetic sentence for any of these situations:

Take-away: I have 5 cookies, and I ate 3. How many are left?

Comparison: I have 5 cookies. You have 3 cookies. How many more cookies do I have than you do?

Doing and Undoing, Staying the Same

Missing addend: I have 3 cookies. How many more do I need to have 5?

Each of these situations can be modeled with cubes or with a number line. Note: Sketch pictures of the cubes and number lines for the following examples.

Take-away Cubes: Start with 5, remove 3, and see what is left.

Take-away Number Line: Move from 0 to 5. Move back 3, and see what point you are at.

Comparison Cubes. Make a stack of 5. Make a stack of 3. Place them next to each other to see what is needed to make the stacks equal.

Comparison Number Line. Mark 5 on the number line. Mark 3 on the number line. How far apart are these two points?

Missing addend Cubes. Make a stack of 3 cubes. Add 1 cube at a time until there are 5 cubes. Count how many cubes were added.

Missing addend Number Line. Move from 0 to 3. Now count how many more jumps you need to land at 5.

If participants are limiting their ideas to a particular kind of story context for 5 − ⁻3, remind them of other types of subtraction situations.

Consider these possibilities for 5 − (⁻3):

Take-away: I have several positive and negative units that represent a net value of 5 positive units. I take away 3 negative units, leaving a number of positive and negative units that represent a net value of 8 positive units.

Comparison: Mark ⁻3 and 5 on the number line. How far apart are these points? This question is not as obvious as it may seem. The distance between two numbers is always a positive quantity. Thus, the distance between ⁻3 and 5 is the same as the distance between 5 and ⁻3, namely 8. However, to model a comparison situation on the number line, "distance from" rather than "distance between" is called upon. "Distance from" implies noting the direction of the movement. Thus the distance from ⁻3 to 5 (expressed as 5 − (⁻3)) is recorded as ⁺8, indicating movement to the right. On the other hand, the distance from 5 to ⁻3 (expressed as ⁻3 − 5) would be recorded as ⁻8, indicating movement to the left.

Missing addend Cubes: Start with ⁻3 and add cubes, 1 at a time, until there is a net value of 5. Count how many cubes you added. On a number line, move from 0 to ⁻3 and count how many jumps of 1 you make to get to 5.

In working to create situations that can model 5 − (⁻3), note that the subtraction sign is associated with a certain meaning and the negative sign is associated with a different meaning. Using the charge model, the subtraction sign is associated with removing certain units, the negative sign is associated with a certain color cube. Using a number line, the subtraction sign is associated with finding the distance from one number to another, and the negative sign is associated with points to the left of zero. The sign of the result indicates if the motion is from left to right or right to left.

Participants should explore both the use of the charge model and the number line as they work. See "Maxine's Journal" (pp. 161–164) for an example of a discussion on using number lines to model subtraction of negative numbers.

In the whole-group discussion, have participants share their ways of representing $5 - ^-3 = 8$. Be sure story contexts, the charge model, and various number-line methods are all displayed. As different forms of representation are offered, ask participants which mathematical ideas the representation makes clear and which ideas are not so clear with that representation. The conversation is not about which of the representations is better but rather recognizing attributes of the various representations.

It is likely this conversation about subtracting negative numbers is going to be confusing for some participants. You might want to assure them that you recognize that not everyone in the seminar will be at the same place of understanding these ideas. Suggest that they think about what part of these ideas makes sense to them. One strategy is for participants to continue to examine the different ideas students offer in Case 25 (lines 720–743), both those that are correct and those that are incorrect. Let them know they can use the writing assignment for next time to continue to think about these ideas and that you will be responding to their writing. This assignment will be a way to continue to work on the math ideas.

Homework and exit cards   (5 minutes)

Whole group

Distribute the "Sixth Homework" Sheet, which involves a problem that addresses a new mathematical topic, the Distributive Property. This topic is one that students work on in the upcoming cases so participants should be made aware that it will be helpful for them to work on this problem before they read the cases.

Exit-card questions:

- What ideas made sense to you in this session? What are you still thinking about?
- What are you learning as you read my responses to your portfolio assignments?

Before the next session ...

In preparation for the next session, read what participants have written about the math they are learning and write a response to each. See the section in "Maxine's Journal" on responding to the fifth homework. Make copies of both the papers and your response for your files before returning the work.

In addition, read over the statements of generalizations your participants wrote to get a sense of how they express these generalizations. You might want to compare this set with the statements you collected in Session 3.

DVD Summary

Session 5: Four Story Problems

First-and-second-grade class with teacher Karen (12 minutes)

The DVD segment shows children discussing problems involving four contexts—fishing, picking flowers, losing balloons, and eating cookies—all using addition or subtraction and the numbers 19, 7, and 26.

It begins as students are discussing the balloon problem: Elizabeth had 26 balloons; 19 of them flew away. How many does she have now?

The teacher, Karen, asks how students can use blocks to represent the word problem. (She has a stack of cubes in her hands, 19 of one color and 7 of another.)

One girl says she doesn't have enough blocks to solve the problem. She needs to have 26.

Karen asks if she is sure that she doesn't have enough.

Others comment that they already know that 19 and 7 make 26.

One student, Meg, takes the cube stack and breaks it right off at 19.

Karen asks how solving the problem about fish and flowers can help them solve the problem about Elizabeth's balloons.

Sawyer says he knows that 7 + 19 = 26. So, he knows that 26 − 19 = 7. Because "when two numbers come together and one goes away, the one that stayed is the answer."

Karen asks what they know about the one that stays.

Hannah says she sees a big relationship about using all three numbers in all of the number sentences. They are in different order. Numbers are being "generous about who goes first."

Karen asks if Hannah's and Sawyer's ideas go together at all. Classmates are silent.

Karen asks a question about the cookie problem.

Sawyer explains that he sees that his ideas match both the balloon and cookie problems.

Math activity and Focus Questions: Addition and subtraction, Two ways to think about the same relationship among quantities

1. The following story problems were given to a class of second graders. Act each one out with cubes, the number line, and/or other representations. How are the problems similar? How are they different? As you work, pay attention to the structure of the problem rather than the numerical strategies that might be used to solve the problem. Write out a generalization(s) for the relationships you are noting.

   - Fish: Peter caught 19 fish in the morning. He caught 7 more in the afternoon. How many fish did he catch altogether?
   - Flowers: Maeve had 7 red flowers. She also had 19 yellow flowers. How many flowers did she have altogether? Can you use what you know from the previous problem to help you with this problem? How?
   - Balloons: Elizabeth had 26 balloons. 19 flew away. How many did she have left? Can you use what you know from the previous problems to help you with this problem? How?
   - Cookies: Codie had 26 cookies. He ate 7 of them. How many cookies does he have left? Can you use what you know from the previous problems to help you with this problem? How?

   Now that you have explored these math problems, turn to the first three cases for this session. (The remaining cases will be discussed later in this session.)

2. In Azar's class (Case 20), many students represent the scale situation with an addition equation. Patrice (line 29) represents it with subtraction, explains how she thought about it, but decides that she must be wrong. Why does Patrice believe that her answer is incorrect? What does Joseph (line 33) understand about the two equations?

3. In Nadine's class (Case 21), some students solve a problem using addition and some solve the same problem using subtraction. In Daisy's class (Case 22), students are working with pairs of related problems. How does their thinking connect with your own work on the four story problems from Question 1? How might Nadine and Daisy use their students' thinking as an opportunity to introduce the idea that the sixth graders in Azar's class are working on?

Focus Questions: Additive and multiplicative identities

1. In Case 23, line 247, Martha says, "I thought that could only happen with 0" and later, in line 305, she asks again, "Is 1 just acting like 0 does?" What is she referring to? What does she mean?

2. In Case 23, line 325, Sharon writes, $1 \times 60 = 0 + 60$. What mathematical ideas does this statement capture? How are those ideas connected to the students' discussion?

3. In Case 24, students continue the discussion from Case 23. In line 514, Mary asks, "If multiplication and addition need different numbers to do that, why do subtraction and division just use the same ones that addition and multiplication use?" What is your response to Mary's question?

Math activity: Subtracting negative numbers

1. We know that 9 + 4 = 13, and the corresponding subtraction equations are 13 − 4 = 9 and 13 − 9 = 4. Although we have studied this relationship by considering the actions of the operations on natural numbers, let us assume that this relationship is maintained when we expand the number system to integers. We know that 8 + ⁻3 = 5. What are the corresponding subtraction equations?

2. Consider our two models for integers, number line and charge. How can these models be used to illustrate subtraction of negative numbers? What story contexts are helpful? What are the challenges that subtracting negative numbers presents?

3. Consider your work on Question 2. How does your work help you interpret the various responses of students in Case 25 (lines 720–743)?

Sixth Homework

Math assignment before reading Chapter 6

Students in Case 26 of Chapter 6 work on the following problem. Solve the problem using a variety of approaches:

Yesterday, I found many flowers in my garden. In the morning, I picked 4 bunches of flowers to give to my family. That afternoon I picked 3 more bunches to give to some friends. Each bunch had 8 flowers. How many flowers did I pick?

Reading assignment: Casebook Chapter 6

Read Chapter 6 in the Casebook, "Multiplying in Clumps," including both the introductory text and Cases 26–30.

Writing assignment: Examples of student thinking

Pose a question to your students related to the ideas in Chapter 4, 5, or 6 of the Casebook. Then think about what happened. What did you expect? Were you surprised? What did you learn? Write a narrative that includes your question, how your students responded, and your reaction to their responses. Include specific examples of student work or dialogue. Examining the work of just a few students, in depth, is very helpful.

At our next session, you will have the chance to share this writing with colleagues. Please bring three copies of your writing to the session.

6

Multiplying in Clumps

Mathematical themes:

- How does the Distributive Property link multiplication and addition?
- Is there a property that links division and addition?
- What are ways to explain what happens when multiplying two negative numbers?

Session Agenda

Sharing student-thinking assignment	Groups of three	25 minutes
Math activity and case discussion: Exploring the Distributive Property	Small groups Whole group	30 minutes 30 minutes
Break		15 minutes
DVD and Math activity: What happens with division?	Whole group Small groups Whole group	15 minutes 15 minutes 15 minutes
Math activity: Multiplying two negative numbers	Small groups Whole group	15 minutes 15 minutes
Homework and exit cards	Whole group	5 minutes

Background Preparation

Read

- the Casebook, Chapter 6
- "Maxine's Journal" for Session 6
- the agenda for Session 6
- the Casebook, Chapter 8: Section 8

Work through

- the Math activity and Focus Questions for Session 6 (p. 210)
- Optional Problem Sheet 3 (p. 211)

Preview

- the DVD segment for Session 6

Post

- Criteria for Representation-Based Proof

Materials

Duplicate

- "Math activity and Focus Questions" for Session 6 (p. 210)
- "Optional Problem Sheet 3" (p. 211)
- "Seventh Homework" (p. 212)
- "The Laws of Arithmetic" (p. 213)

Obtain

- DVD player
- index cards
- cubes

Examining the Distributive Property

Many people approach multiplication problems by breaking the numbers apart; for instance, one way to solve 28×6 is to rewrite it as $(25 \times 6) + (3 \times 6)$, calculate each of the products, 150 and 18, and then combine them to get 168. The mathematical principle that underlies this process is the Distributive Property. That is, $(25 + 3) \times 6 = (25 \times 6) + (3 \times 6)$. In its general form, this property is stated as $(a + b) \times c = (a \times c) + (b \times c)$. In this session, participants will use story situations and diagrams to understand why this property makes sense and how it is applied.

Consider this story context for $28 \times 6 = (25 \times 6) + (3 \times 6)$: I want to give 28 students each 6 pencils; how many pencils will I need? This context, which is modeled by the expression on the left side of the equation, can be slightly elaborated to fit the expression on the right side: Of the 28 students, 25 are boys and 3 are girls. First, find the number of pencils given to the 25 boys (25×6) and then, find the number of pencils given to the girls (3×6). In both expressions, there are the same number of students, the same number of pencils per student, and the same number of total pencils. More generally, if a boys and b girls are each to get c pencils, you can think of the number of pencils needed as $(a + b) \times c$ (multiply the total number of children by the number of pencils each child gets) or as $(a \times c) + (b \times c)$ (the number of pencils needed for the boys added to the number of pencils needed for the girls).

Multiplication can also be represented by using arrays or the area of rectangles. Consider the diagram below. The entire rectangle represents the expression 28×6. Focusing on the two smaller rectangles leads to the expression $(25 \times 6) + (3 \times 6)$. Because the area of the larger rectangle (28×6) is found by adding the areas of the two smaller rectangles, then $28 \times 6 = (25 \times 6) + (3 \times 6)$.

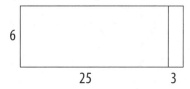

To represent the generalization, the rectangle might be labeled as follows:

The area of the large rectangle can be viewed as the product of the lengths of the sides—$(a + b) \times c$—or the sum of areas of the two smaller rectangles—$(a \times c) + (b \times c)$.

Similar arguments can be made with stacks of cubes or by using array cards.

Maxine's Journal

November 12

Last night, the main focus of our work was on the Distributive Property. Although the idea of the Distributive Property is not new, it was so interesting to watch everyone dig in deeply to explore the operation of multiplication and what is involved in "multiplying in clumps." We also spent time examining division to consider whether a similar property holds.

Then after working to understand the property more deeply, we looked back at negative numbers. Specifically, we considered two different ways of thinking about why the product of two negative numbers is positive. One argument is based on the Distributive Property and consistency of the number system and the other argument is based on contexts that accommodate integers and embody the operation of multiplication.

Before this work began, participants had a chance to share the student-thinking assignments they had prepared for the session.

Sharing student-thinking assignment

For their student-thinking assignment, most participants had done something with the Distributive Property. In fact, quite a few gave their students the flower problem from Lucy's case: She picked 4 bunches of 8 flowers in the morning and 3 bunches of 8 flowers in the afternoon; how many flowers did she pick that day?

Participants shared their students' different approaches to solving the problem. As they discussed the student work, I frequently heard, "I wish I had…" or "If I were to do this again, I would…" I asked the participants to write these thoughts on their papers so they would remember them and also for me to acknowledge their ideas in my response to their writing.

Math activity and discussion: Exploring the Distributive Property

For most of the remainder of the session, the math activity and case discussions were intertwined. There were three main ideas I wanted to work on: exploring the Distributive Property, considering whether there is an analogous rule for division, and making sense of multiplying negative numbers.

Small groups

Participants began by working on Part 1 of the Math activity and Focus Questions sheet, "Exploring the Distributive Property." As I observed the groups, I recognized that most participants were working to create contexts that would help them think about the relationships.

Phuong said, "Let's say you have to carry books to school; each of 28 students gets 6 books—that is 28 × 6. The books are heavy and you are unable to carry them all at once. On your first trip, you carry the books for 10 students. On your second trip, you carry books for another 10 students. Then on the third trip, you carry books for the remaining 8 students."

Phuong had drawn the following diagram: The entire rectangle represents all the books you need to take to school, broken into the amounts you carry in each of the three trips.

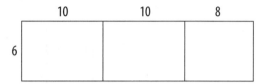

M'Leah, looking at Phuong's diagram said, "You could see them as three classrooms, or you could look at the area of the floor altogether as 28 × 6."

Antonia suggested, "You could put the walls in differently to make an auditorium that is 25 by 6 and a storage closet that is 3 by 6. All that is the same as 28 by 6."

When I stopped by Charlotte's group, she told me that she had spent a lot of time looking at the equation written in the introduction—$(a + b) × c = (a × c) + (b × c)$—and was still having trouble making sense of it. "When I look at those parentheses, it means 'do this first,' doesn't it? But, I don't see that there's anything I can do."

I suggested to Charlotte that she set aside the algebraic notation for a little longer. I told her that I understood she really wanted to figure it out, and we would get back to it in our whole-group discussion. It might make more sense to her after she creates diagrams and contexts to help her think through the ideas. For example, she might examine the various representations of 19 × 4 offered by Carl's students in Case 29.

Kaneesha, Leeann, and Mishal were looking at the representations in Carl's case and were intrigued by Chandler's thinking—Strategy 1. Chandler converted the multiplication problem, 19 × 4, to addition; decomposed each of the 19s to 10 + 9; rearranged all the addends to group the 10s together and the 9s together; and then converted back to multiplication: (10 × 4) + (9 × 4). Mishal was saying, "It just looks so neat and logical." The group then went on to examine the rest of the strategies offered by the class. Kaneesha said, "Once I saw what Chandler did, I can think of the rest of the representations in that same way. Whether we're looking at circles with tally marks in them or some kind of array, you break apart the 19 into 10 + 9 and then change the way things are grouped. The representations are set up so you don't actually have to move anything, you just change the way you look at it."

Grace, Madelyn, and James were discussing Simone's story problem: "Bob buys 4 boxes of doughnuts with 19 inside each box. Larry buys 4 boxes of doughnuts with 10 in each box and 9 boxes of doughnuts with 4 doughnuts in each box." Grace had drawn a representation in the shape of an *L*. She explained that this represents the two kinds of doughnuts Larry bought. The top array shows 4 across for the 4 boxes, with 10 doughnuts in each. The bottom shows 9 across for the 9 boxes, with 4 doughnuts in each. She said, "We know that Larry and Bob bought the same number of doughnuts, but that doesn't help us see why the Distributive Property always works out."

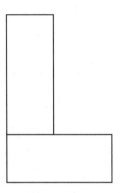

James said, "Well, maybe we can just rotate one of those arrays. Then you get the array with 19 × 4."

Grace said, "But then it no longer describes what Bob bought. When you look at the vertical dimension, does it stand for 4 boxes or 4 doughnuts? It changes."

Madelyn said, "You know, the problem here is pretty subtle. I can understand why Carl decided not to sort it out in the moment."

By now, many groups were thinking hard about (10 + 2) × (10 + 8), and I suggested that everyone start working on the rest of the problems if they had not done so already. Then after a few more minutes, I decided to bring participants together as a whole group. As I tried to call for attention, the small-group discussions continued. The participants were deeply involved in their small-group discussions, but I truly believed they would get more from a whole-group conversation. Additionally, there were other issues I wanted to get to in this session, so I persevered and pulled us all together.

Whole group

Before beginning, I wrote on the board:

$$12 \times 18 \qquad (10 \times 10) + (2 \times 8)$$

I began by pointing out that everyone had done considerable work on this problem. "What pictures, representations, and stories did you use to explain why $(10 \times 10) + (2 \times 8)$ is not adequate to solve 12×18, and what is adequate? Does someone have a story or a picture that really worked for you?"

Yanique spoke up first, "I was thinking about students and pencils: 10 students get 10 pencils each, and that is 10×10."

Yanique had a good idea, and I knew from my observations of her small-group discussion that she had thought the whole issue through, but she was not presenting it all to the group. I asked her to explain the whole problem.

Yanique then took us through the ideas. "You have 12 students and they each get 18 pencils. If you take 10×10, that stands for 10 students each getting 10 pencils; 2×8 is 2 more students each getting 8 pencils. You have 12 students, but none of them got the full 18 pencils."

Lorraine, who had worked in Yanique's group, said, "That made sense to me. I have passed out pencils to students; some have 10 and some have 8; and so that shows why it's wrong."

I asked, "So did you go further to show how to fix this?"

Lorraine responded, "I can tell the parts that are missing. Those 10 students need 8 more pencils. The other 2 students need to get 10 more pencils. So, that's 10×8 and 2×10."

According to Lorraine's instructions, I added to the expression on the board, so it now read:

$$12 \times 18 \qquad (10 \times 10) + (2 \times 8) + (10 \times 8) + (2 \times 10)$$

Jorge pointed out that the expression is now equal to 12×18, so I could put in the equal sign.

$$12 \times 18 = (10 \times 10) + (2 \times 8) + (10 \times 8) + (2 \times 10)$$

Everyone agreed that this equation was correct, so I asked if anyone had a different way to look at it.

Risa said her group made two pictures, and she came up to draw them.

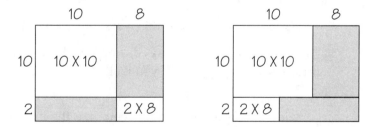

I asked how Risa's pictures connect to the pencil story, and Denise talked us through it. To make sure everyone could follow, I repeated Denise's

explanation, pointing to the sections in both pictures that show the 10 stu- | 125
dents who first received 10 pencils and the 2 students who first received 8.
The gray areas are for when the 10 students received 8 more pencils and the
2 students received 10 more pencils.

M'Leah said that, in her group, they drew a picture like Risa's first one | 130
and had a long discussion about it. "No matter where you move that vertical
line, the number of pencils you need to buy is the same. You can move that
line—like make it 9 and 9—and the picture would be the same, just with that
line moved over."

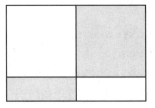

M'Leah continued. "You could also move the horizontal line, and it would be | 135
the same."

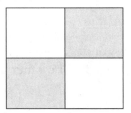

"But if you take pieces out, that's when it's not the same."

So far, I had heard from Yanique, Lorraine, Risa, and M'Leah, and they
seemed to be clear about what is going on with this problem. I suspected there
were more insights to be gleaned, so I stayed with it. "Did anyone have a way
to think about this problem that was different from these two?" | 140

James said, "We did skip-counting."

When I asked for more information, he explained that they thought about the
first problem, 28 × 6, as skip-counting. "You can count 28, 56, 84, 112, 140, 168.
That is counting by 28, or you can count by 25—25, 50, 75, 100, 125, 150—and
then count by 3—3, 6, 9, 12, 15, 18. Add 150 to 18 and you come up with 168." | 145

I asked, "So why should this work? Why does this make sense to you?"

James and his group paused, not sure how to respond to the question.
Lorraine spoke up again. "If you look at it on an array, you would have 25 × 6
and 3 × 6."

I said, "So, Lorraine, does it seem like the image of the array makes the | 150
logic apparent, more than when you do the skip-counting?"

Lorraine nodded, but Phuong had another idea. "You can see the pattern
with skip-counting, too. When you count by 25, you are losing 3 for each

jump, so you end up with 18 less than what it should be. When you count by
3, you make up for all the 3s you lost." 155

I pointed out that you could relate this problem to a story context. Lourdes
said that her group talked about pencils, too. Six students each received 25
pencils on Monday and then on Tuesday, they received the other 3."

At this point, I realized that, although the Distributive Property is named
in the introduction to the Casebook, we had not yet said it out loud in the ses- 160
sion. "Today we've been working on the Distributive Property. That is, when
you multiply two numbers, you can break one up, multiply each part by the
other number, and add the products. In algebraic notation, it is represented
this way." I wrote on the board

$$a \times (b + c) = (a \times b) + (a \times c)$$ 165

The term *Distributive Property* took M'Leah back to her high school education. "When I was in high school taking algebra, and we had to do $(a + b) \times
(c + d)$, our teacher showed us how to make a happy face."

I had no idea what M'Leah was talking about, so she came up to the board
to show us. 170

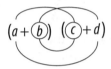

As we all looked at M'Leah's picture and found the face, we started to
laugh. She had circled b and c to show eyes; the arc linking b and c is a smile;
the arc linking a and d is the chin, and so on. M'Leah elaborated, "Once you
make the happy face, you multiply the numbers or letters that are linked and
add them all up." 175

So that everyone could see the equation M'Leah was describing, I wrote,

$$(a + b) \times (c + d) = (a \times c) + (a \times d) + (b \times c) + (b \times d).$$

I pointed out that this method involved repeated application of the
Distributive Property. The statement of the property itself is that one factor
remains fixed and the other is decomposed into two addends. 180

Referring back to M'Leah's comments, Risa said, "When I was in school,
we did not use a happy face, but we talked about the FOIL method."

There were murmurs of acknowledgment in the room. Actually, when I
was in school, my teachers had never taught me the mnemonic, FOIL, but I
had learned it from my students years ago. The letters of *FOIL* tell you which 185
terms to multiply—first (a and c), outer (a and d), inner (b and c), last (b and
d). Risa came up and explained this to the group.

I was feeling a little uneasy with all this talk about how to remember steps
that seemed completely disconnected from mathematical reasoning. The

participants were enjoying this too much for my taste. Then M'Leah laughed and declared, "That happy face made it so I could get the right answers on my tests—I did get good grades—but I never understood what it was all about." 190

Yanique announced, "I always got good grades, too. But this class is making me understand algebra for the first time."

Needless to say, I was relieved to hear those remarks. 195

I said, "A lot of teachers give students these ways to remember the rule because so many students say $(a + b) \times (c + d)$ is the same as $ac \times bd$. It's like what we were investigating before. Some students will first think they can solve 12×18 by using $(10 \times 10) + (2 \times 8)$. So my question is, if students are making this error in algebra class with a, b, c, d, did they ever have a chance to think through what's going on with multiplication and multi-digit numbers?" 200

That was a rhetorical question, so I went on. I now wanted the group to apply the thinking they had done for 12×18 to interpret $(a + b) \times (c + d)$. Pointing to the equation I had written after M'Leah drew her face, I said, "Let's see if we can apply our mental images to this equation. Let's think of a story about distributing pencils to students." 205

Within 2 minutes, the group came up with the following context: There are students in a class on the first floor (a) and students in another class on the second floor (b), and they all receive some pencils on Monday (c) and some pencils on Tuesday (d). 210

Antonia said, "So $a + b$ represents students altogether, and $c + d$ represents the number of pencils each student receives."

I acknowledged what Antonia said, and then took the group through each part of the equation. Antonia explained the first half—$(a + b) \times (c + d)$ gives us the number of pencils that are distributed altogether. So we looked at each term on the other side of the equation: $a \times c$ is the number of pencils the first-floor class receives on Monday; $a \times d$ is the number of pencils those same students receive on Tuesday; $b \times c$ is the number of pencils the second-floor class receives on Monday; $b \times d$ is the number of pencils those students receive on Tuesday. Add up all those terms, and you have the number of pencils that are distributed altogether. 215
220

Charlotte said, "So, up until now, whenever I have looked at an equation like that, a part of my brain would freeze. I had learned that the parentheses tell me I am supposed to do something first, but when I see an equation like that, there is nothing to do. Now, I'm rethinking this. It's not about saying what to do first, but it's a way to restate what something is. You can look at the parts together or separate the parts." 225

Charlotte was saying something profound, and I told her so. "When you look at '$(a + b)$,' you can see it as an object that you can bring meaning to, as opposed to a set of rules that you are supposed to enact. Using our example, $(a + b)$ stands for the number of students in the two classes." 230

After a pause, Jorge brought us back to why $(a + b)(c + d)$ is not just $ab + cd$. "You can see it in a diagram, too."

	a	b
c	a × c	b × c
d	a × d	b × d

Now, with this area diagram, we went through each part and showed where it appeared in the equation and how it related to the story about two classes of students who received pencils on Monday and Tuesday.

James said, "I like the area diagram."

Kaneesha said, "I like the story situation."

Madelyn said, "I am thinking about what we've been working on in the last few sessions, like subtracting a negative and the Distributive Property. We're continually looking at ways to represent the situation, and we keep looking for different ways. This really helps me deepen my knowledge about an idea."

Lorraine agreed. "Yeah, before I didn't do much work with models, but I'm doing more and more of it with my students, and it seems to make a difference in their thinking."

Leeann said, "It's not just about the different ways, but seeing how the different ways are related."

Wow. I was really pleased to hear that from Leeann. I concluded the discussion by picking up on what Leeann said. "I think Leeann is highlighting an important idea. It's not just about creating a variety of representations but also about comparing and connecting the representations. Seeing how they fit together makes more insights available to us. By seeing differences in what is revealed by several models, we come to understand things that would be difficult to see with only one model. We notice more because differences become more apparent."

Charlotte added, "Looking at all the different representations helps me not only understand the Distributive Property more deeply, but for the first time, I am seeing an inkling of meaning in algebraic notation."

DVD and Math activity: What happens with division?

We began the next discussion with a DVD clip that shows students working on the following problem: 198 students need to be seated in the cafeteria; tables seat 14 students each. How many tables are needed?

The fourth-grade girls in the DVD clip first find the number of students seated at 10 tables, then the number seated at 4, and then they show they still need an additional table for 2 students. At the end, the girls' work looks like this:

$$14 \times 10 = 140$$
$$14 \times 4 = \underline{56}$$
$$196 \text{ with a remainder of } 2$$

After showing the clip I asked, "These students were solving a division problem. What do you notice?"

Jorge opened the discussion. "I noticed that the teacher asked them questions to connect what they had done with their calculations to their representations. In order to solve the division problem, they thought about multiplication."

James added, "The teacher had them outline the 10 tables with 14 at each table, and then show the other parts. I see the Distributive Property at work here, but do students?"

I commented, "You are asking a good question: Do students see the Distributive Property at work here? I want to make sure we all recognize the Distributive Property here. What is it you see?"

Madelyn took that on, "You can say $(14 \times 10) + (14 \times 4) = 14 \times 14$. As Jorge said, they solved the division problem with multiplication."

I commented, "Although this is not what the girls did, what do you think about this?" I wrote out:

$$198 \div 14 = (140 \div 14) + (58 \div 14)$$

I continued, "Although the girls thought of it as multiplication, they were asking how many tables were needed to seat 198 students. They first determined how many tables seat 140 students, and then they found how many tables seat 58 students, and then they added the results. It looks pretty much like the Distributive Property, but it's division instead of multiplication. It seems to work here, right?"

Most participants nodded, or indicated some kind of agreement. So, I asked them to dig into this question more deeply. I asked the group to look at Question 5 on the "Math activity and Focus Questions" Sheet to examine the following pairs of expressions.

a. $(12 + 4) \times 2$ $(12 \times 2) + (4 \times 2)$

b. $(12 + 4) \div 2$ $(12 \div 2) + (4 \div 2)$

c. $12 \times (4 + 2)$ $(12 \times 4) + (12 \times 2)$

d. $12 \div (4 + 2)$ $(12 \div 4) + (12 \div 2)$

Before the group got started, I said, "When we look at these examples, they look very similar. However, there is something that's quite different with one of them. I would like you to figure out what's going on there."

The first and third pairs are simply examples of the Distributive Property of Multiplication over Addition, and so the expressions are equal, but the second and fourth expressions involve division. Yet as participants looked more closely, they realized that the second pair of expressions is equal, but the fourth pair is not.

I gave participants about 10 minutes to work in small groups before I brought them back to the whole group. I wanted to think through this situation together.

When I first asked if someone would describe what is happening here, Madelyn said, "'Groupiness' is staying constant in the first three. In the last one, the groupiness feature is not held to."

I asked what she meant by *groupiness*, and she explained further, "In the first pair of expressions, you have 16 groups of 2 written out in two ways. In the second pair, you still have groups of 2 in both expressions. But in the fourth pair, you start with groups of 6, but then you have groups of 4 and groups of 2."

I asked if that wasn't true of the third pair of expressions as well. Madelyn looked at that and said, "Well, yes. But you can also think of it as 6 groups of 12 written two ways, or you can say, you have 12 groups of 4 and 12 groups of 2, so when you match up the groups, you have groups of 6 again. That doesn't happen with the fourth pair. There you end up with 3 groups of 4 and 6 groups of 2."

I could see that Madelyn was thinking about the meaning of multiplication and division as she described the differences among the various expressions. However, I was not so sure if anyone else was following. So, I turned to the rest of the participants and suggested we come up with a story context to think about this.

Risa said, "Let's say we're distributing treasures."

When I asked her to apply that context to the second pair of expressions, she continued, "You have 16 treasures to be distributed between 2 people."

I commented, "OK, so we're not talking about groups of 2. We're thinking of distributing the treasures into 2 groups."

Lorraine added, "The treasures are 12 opals and 4 diamonds. So the expression on the right says you first distribute the 12 opals and then the 4 diamonds. Each person still receives the 8 treasures."

I summarized, "OK, so we have a story context that assures us that the two expressions in part *b* are equivalent. What about part *d*?"

M'Leah said, "Now you have 12 treasures and you have 6 people to share them with. You have 4 friends in one room and 2 friends in the other. So each person receives 2 treasures. That fits the expression on the left, but the expression on the right says that you have 12 treasures to share among 4 friends and another 12 treasures to share between 2 friends. It's a different situation. The expression on the right describes a situation in which you have a lot more treasures but the same number of people, so it's different."

As I looked around the group, almost everyone was nodding, feeling very satisfied that, together, we had come up with a way of thinking about what was happening in part *d*. Only Charlotte was frowning. When I asked her to speak up, she said, "I get it, that the two expressions in part *d* don't go with the same situation, but it sort of bothers me that the Distributive Property works some of the time but not all the time. Why does it work for part *b*?"

I was not exactly sure what Charlotte was saying, but I decided to offer a different way of looking at it. I said, "As we've said before, division can be thought of as multiplying by the reciprocal. So let's take a look at what happens in part *b* if, instead of dividing by 2, we multiply by $\frac{1}{2}$." Then I wrote:

$$(12 + 4) \times \frac{1}{2} \qquad (12 \times \frac{1}{2}) + (4 \times \frac{1}{2})$$

I continued, "When we do that, it becomes an application of the Distributive Property of Multiplication over Addition. Let's take a look at what happens when we do the same for part *d*."

$$12 \div (4 + 2) = 12 \times \frac{1}{6}$$
$$(12 \div 4) + (12 \div 2) = (12 \times \frac{1}{4}) + (12 \times \frac{1}{2})$$

Jorge said, "Now, if you apply the Distributive Property to the second expression, you get $12 \times \frac{3}{4}$, not $12 \times \frac{1}{6}$."

Charlotte was nodding, "OK, I can see that."

Math activity: Multiplying two negative numbers

There was one more major mathematical idea that I wanted to address in the session. In Case 30, Carl's students present ways of thinking about why the product of two negative numbers is positive. One student, Ian, found a proof on the Internet based on the Distributive Property. Although Ian's proof used specific numbers, it approaches an algebraic argument. (In fact, if he had used *a*, *b*, and *c*, instead of 3, 4, and 12, he would have the proof of the generalization.) I wanted participants to think about Ian's argument, together with the other suggestions offered by Carl's students, and consider the different ways of knowing that these different arguments present.

I gave participants some time to work in small groups to discuss lines 580–606. Then I brought them back together. I began the whole-group discussion by asking if someone would explain Ian's method.

James came to the board and presented the argument: "Ian knows that $0 \times n = 0$. So he replaced *n* with ⁻4 and 0 with 3 + (⁻3)." He wrote:

$$(3 + {}^-3) \times ({}^-4) = 0$$
$$(3 \times {}^-4) + ({}^-3 \times {}^-4) = 0$$

James continued, "He just used the Distributive Property. So now we know that $3 \times (^-4) = {^-}12$." 380

I stopped James there and said, "You know, we haven't discussed that before. How do we know that is true?"

Lorraine spoke up, "We can think of $3 \times (^-4)$ as $^-4 + {^-}4 + {^-}4$, which equals $^-12$."

I said, "So, when one factor is positive, we can still rely on repeated addition." 385

Then I nodded at James, and he went on, "So, we have $^-12$ plus something equals 0, so that something must be 12."

James then wrote under the first two equations,

$$^-12 + 12 = 0$$
$$(^-3) \times (^-4) = 12$$

390

Madelyn said, "I really like that argument. It feels so satisfying to have everything fall into place that way."

I asked about other participants' reactions.

Grace said, "It makes me feel like, why didn't I think of that?"

M'Leah said, "There's something that just feels like a trick to me. This is really hard to digest. Like, why do you have to bring in 0? What if you come up with something equal to another number? I will need to think about this some more." 395

June asked, "So, Ian is doing this because he wants to show that $^-3 \times {^-}4 = 12$? That is why he went through all those steps?" 400

I nodded and June continued, "I understand what's going on; I can follow each step. But, this type of argument doesn't work for me. It's not convincing. For me, I need imagery."

I asked, "What about Cindy and Nadir's arguments? Did either of those arguments make sense to you?" 405

June said, "I liked Cindy's argument with the hot air balloon and the sandbags. I can picture that image—each bag stands for $^-1$, and if you throw over several groups of sand bags, the balloon goes up. I get that other people won't be convinced by the balloon, but that's what convinces me."

Mishal said that she still did not understand the hot air balloon and asked for an explanation. I quickly went through the context: You have a hot air balloon hovering in mid-air. When you give it a blast of hot air, the balloon goes up 1 foot. If you release hot air, the balloon goes down. That is like adding and subtracting a positive amount. If you add a sandbag, the balloon goes down; if you throw out a sandbag, the balloon goes up. That is like adding and subtracting a negative amount. When Cindy used the balloon context to think about multiplying negative numbers, she thought of throwing 4 groups 410 415 of 5 sandbags over the edge. Each group stands for ⁻5; throwing over 4 groups is like multiplying by ⁻4.

Mishal nodded and said, "OK. I can picture that."

Lorraine said, "I like the story about removing debt. If you remove 3 debts and each debt is $4, you end up $12 ahead of where you were."

I now emphasized the point about different ways of "knowing." "Let's pause for a moment not to make a judgment on the different arguments, but to acknowledge that there are different ways of coming to know a piece of mathematics. One way is to rely on the consistency of the number system. We talked about this in Session 5. At that time, we were looking at how addition and subtraction were related. If $a + b = c$, then $c - a = b$ and $c - b = a$. In Session 5, we said that because $13 + (⁻8) = 5$, we can conclude that $5 - (⁻8) = 13$. Some people might have been satisfied with that, but then we went further to think about different models and contexts to help us bring more meaning to the idea that subtracting a negative number is equivalent to adding a positive. We might say that all the work we did with models and contexts was helping us to cultivate mathematical intuition.

"So today we've been thinking a bit about multiplying negative numbers and find that the product of two negative numbers must be positive. We can come to that through the consistency of the number system. If we accept that the product of anything times 0 is 0, understand the Distributive Property, and recognize that any number plus its additive inverse is 0, then we must conclude that the product of two negative numbers is positive. Then, just as we did with subtracting negatives, we can also think about models and contexts that fit this finding. We might do that to help us cultivate intuitions about how these numbers behave under the different operations, including multiplication.

"But, let me say a bit more about Ian's argument. If the Distributive Property is not yet solid for you, or if there is any step that you haven't already thought through, then the argument makes no sense. If we move into this territory with our students, we don't want them to be completely befuddled or to rely exclusively on memorizing rules that have no meaning. We need to pay attention to whether they need to refer back to models and contexts and that's why we're doing so much work with models and contexts in RAO."

Homework and exit cards

I distributed the "Seventh Homework" Sheet together with a page titled "The Laws of Arithmetic" and pointed out that for the next session, there were two writing assignments. First, I asked that participants reflect on how their ideas about teaching and learning mathematics have been influenced by this seminar. Second, I asked that they look over the Laws of Arithmetic and write how the ideas connect with the work they have been doing in the seminar.

Multiplying in Clumps

I pointed out that, in this context, we can use the words *laws* and *properties* interchangeably.

For their exit cards, I asked participants to write about a new mathematical idea that they gained from today's session. Many of them wrote about how to decompose numbers in a division problem. Phuong wrote explicitly about when it works and when it does not. "Division is different compared to the use of the Distributive Property in multiplication. Only the dividend can be broken up—not the divisor, which tells us either the number of groups or group size the 'things' are being put into. The meaning of division (dividing or sharing in equal groups) is lost if the divisor is changed."

Grace, Kaneesha, Charlotte, M'Leah, and several others indicated that they realize that sometimes it works to decompose the number in a division problem and sometimes it does not, but they still need to think about when it works and why.

Leeann said that she can finally see the power of an array representation, but it was necessary to have that connected to pictures and stories. "In our third-grade math book, they encourage us to teach students to create and use arrays. I was never convinced that they can be helpful. After tonight's session with the arrays, I see how they can be used, but now I know I have to use pictures or drawings first. I think that if I can relate drawings to arrays, students will understand them better."

Risa said that she has been working on fractions with her students, and this session helped her look at multiplying mixed numbers in a new way. "As we've been looking at the Distributive Property, I have been coming back to my work with my fifth graders on fractions and mixed numbers. I have been quickly reviewing the four operations with fractions, so I did not dwell on or go into a lot of discussion due to time constraints. But I may still pose the question to my students: Does $2\frac{1}{2} \times 3\frac{1}{4} = (2 \times 3) + (\frac{1}{2} \times \frac{1}{4})$?"

David and Madelyn wrote about how cool it was to see Ian's explanation for why the product of two negatives is positive.

Lorraine said that she was so pleased to be able to think of a context that helps her make sense of multiplying negative numbers.

Responding to the sixth homework

November 14

By now, everyone in the group is writing substantial pieces for their student-thinking assignments. In each piece of writing, there is quite a bit to think about and quite a bit to respond to. I would like to share Vera's assignment.

Vera is a principal and in the beginning of the seminar, she indicated that she is taking the course so that she could talk to the chair of the high-school mathematics department about teaching algebra. For that reason, I especially wanted to help her find language to describe the work we are doing, and a few sessions back I gave her an article, "Reasoning about Operations: Early Algebraic Thinking, Grades K through 6," written by Deborah Schifter (an author of the module), from the 1999 NCTM Yearbook, *Developing Mathematical Reasoning, Grades K through 12*. After she wrote her student-thinking assignment, she responded to that article.

Vera

As I read through Chapter 6 and worked through the problem about bunches of flowers, I was struck by the fact that several years ago when I took BST, I, too, had difficulty understanding why I could not take numbers apart in multiplication in the same way I did for addition. Soon I learned about arrays and multiplying in clumps.

This prompted me to try a similar multiplication problem with two students, Marcella and Lionel.

The problem I posed:

Last weekend I went apple picking. In the morning, I picked 4 bags of apples with 8 apples in each bag. In the afternoon, I picked 3 bags of apples with 8 apples in each bag. How many apples did I pick altogether that day?

Marcella's work: morning $8 \times 4 = 32$
                     afternoon $8 \times 3 = 24$

$8 \times 7 = 56$ *answer*
or $32 + 24 = 56$

the answer is 56 apples

Lionel's work:

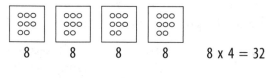

$8 \times 4 = 32$

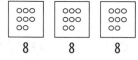

$8 \times 3 = 24$

Lionel first tried adding all the 8s, then tried to multiply 32×24 and said, "No, you have to add it together."

I then asked each one if he or she could do the problem a different way.

Marcella's way: You count by 8 to 7 bags.
        8, 16, 24, 32, 40, 48, 56

Lionel could only think of adding: $8 + 8 + 8 + 8 = 32$
$$8 + 8 + 8 = \underline{24}$$
$$56$$

Marcella added the 4 and the 3 and then thought about 7×8.

Lionel first thought about adding the apples in the bags and then moved to $8 \times 4 = 32$; $3 \times 8 = 24$; *then* $32 + 24 = 56$.

Lionel did not like Marcella's way of skip-counting, yet he was willing to add all the 8s.

I asked if either knew about arrays and if they could draw that.

Both looked puzzled, but each could articulate 7 bags with a group of 8 apples in each bag.

I then followed by writing: $(4 \times 8) + (3 \times 8)$.

Marcella quickly said, "4 × 8 + 3 × 8. Now, 32 + 24 = 56."

Then I asked what Lionel would write after the equal sign

$$4 \times 8 + 3 \times 8 = ?$$

Lionel wrote: $(4 \times 8) + (3 \times 8) = 7 \times 8$

I asked Marcella to read it.

MARCELLA:	*Four times eight, plus three times eight equals seven times eight.*
TEACHER:	*Is that true?*
MARCELLA:	*I guess so, because both sides equal 56, and 56 is the answer.*

I was not sure if either student would be able to attack this multi-step problem, but both surprised me and did so with ease; Marcella with numbers and Lionel with pictures.

Both M and L could visualize the problem and solved it with multiplication and addition. M: $(4 + 3 = 7$ bags$) \times 8$ apples
$$\text{L: } 4 \times 8 = 32, 3 \times 8 = 24, 24 + 32 = 56$$

I was surprised that neither student could relate to the word "array" but was pleased to see Marcella count by 8s.

Because I have a very limited understanding of the Distributive Property, I could not tell if either student understood it that way.

TO MAXINE: Thank you for sharing the article, "Reasoning about Operations: Early Algebraic Thinking, Grades K through 6."

As I read through the article I immediately connected with the opening paragraph about the confusions seen today among secondary students studying algebra and about the difficulty secondary math teachers have tapping into what students already know about number operations.

After reading the article and participating in this RAO seminar, I am rethink-
ing some of my original beliefs about early algebra as equality, representation,
variables, and so on. I am now thinking much more about how elementary-
age students engage in, and think about, the four operations of addition,
subtraction, multiplication, and division.

I am also trying to fully understand the ideas of the Commutative, Associative,
and Distributive Properties. You have been very generous with helping me to
understand these concepts and how students might engage in these ideas through
switch-arounds, negative numbers, multiplying in clumps, the size of numbers,
and the nature of the questions.

You have given me a whole new lens to use when observing math lessons in
our elementary classrooms. In addition, I have a greater appreciation for each
student's level of understanding about operations and how each operation
functions with whole numbers, negative integers, fractions, and with bigger
and bigger numbers.

It also helps me think about how much an elementary math teacher needs to
know; how different models help demonstrate and extend differing teaching
points and extensions of a problem; and how much I have to support continu-
ous professional development for all of our staff, both novice and experienced.

Truly, I am becoming transformed in my own thinking and I agree that the
hope is to send our elementary students on to middle school and high school
with a better foundation and understanding about algebraic concepts.

Thank you for that, and thank you for tapping into and supporting my ability
to understand mathematics. You have been very generous with your feedback
and with your time to make this course beneficial for me and for the many
math coaches and teachers in the room.

Dear Vera,

That is really interesting that you recognize your own thinking about how
multiplication and addition have evolved. Sometimes, once we do understand
something, it is hard to remember how we thought about it differently at one
time.

Marcella seems very confident with multiplication. She seemed able to think
with the numbers. Lionel immediately drew a picture in which he can see
every single apple. It was interesting to me that he did not just draw bags
and label them with an "8," but that he put in all the individual circles, too.
I wondered why he changed his mind about multiplying 32 and 24. I would
have been interested in asking him why he decided to add the numbers. Did
he just see that multiplying 32 and 24 would be too difficult or that it would
give him too large a number? Do you think he could explain why he would
add the two numbers?

I was also curious why Lionel did not like Marcella's skip-counting. From your
point of view, he is doing something very similar when he adds the 8s, but he sees

that as quite different. I wonder if there would be some way for him to see how 605
adding 8s and skip-counting are similar.

I am fascinated by the fact that Lionel wrote (4 × 8) + (3 × 8) = 7 × 8. So often,
students assume that the equal sign indicates that they should write the answer
to the problem. I would be interested in why Lionel wrote 7 × 8 and how he
would explain what he means. Marcella seemed to accept that the two expres- 610
sions on either side of the equal sign are equal, and she explains this very well.
How would Lionel explain this?

Now, I am interested in what you mean about your limited understanding
of the Distributive Property. It seems to me that these students, especially
Marcella, understand some important aspects of the Distributive Property. 615
Marcella writes that "8 × 7 = 56" or "32 + 24 = 56" gives the answer to the
problem. It seems to me that she is seeing that the two parts, 8 × 4 and 8 × 3,
can be added together to give the solution to 8 × 7. So, is she not applying the
Distributive Property? Do you think she (or Lionel) just see 8 × 7 as another
way of writing the problem, but do not see how it is related to (8 × 4) + (8 × 3)? 620
When Lionel wrote (4 × 8) + (3 × 8) = 7 × 8 at the end of the lesson, did you
think he was seeing that 7 × 8 can be broken up into two parts by breaking
up the 7? What would happen if you started with a problem like 6 × 9—with
numbers that are not too big, but big enough so they do not "just know" the
answer—and asked them to break it up into smaller problems? 625

I was also interested in your comments about the article and the seminar. I
am so glad that you feel your understanding of the operations is expanding
and that you can use these ideas in your work.

Maxine

Detailed Agenda

Sharing student-thinking assignment (25 minutes)

Groups of three

The first activity of this session is designed to provide the participants an opportunity to share their reflections about the mathematical thinking of their students. Participants should read all episodes before beginning their discussion so that they may include particulars of a single case, as well as commonalities among all of the cases. Remind participants that they are responsible for organizing their discussion to ensure that each paper receives attention.

Math activity and case discussion: Exploring the Distributive Property (60 minutes)

Small groups (30 minutes)
Whole group (30 minutes)

During this time period, participants work on Part 1 of the "Math activity and Focus Questions" Sheet, which provides an opportunity for participants to examine the way the Distributive Property is used in computational strategies for multiplication. This work builds on the mathematics of Session 5 in *Building a System of Tens*. Before participants begin the small-group work, you might want to briefly review the term *Distributive Property*, which is referred to in the introduction to the cases and also in the title of the "Math activity and Focus Questions" Sheet. Introduce this activity as an opportunity for participants to explore the Distributive Property more deeply.

Participants should approach this work using diagrams and story contexts, as well as numbers and symbols. If some groups are having difficulty exploring the questions with a variety of representational tools, you might interrupt the small-group work for a few minutes to review a variety of ways that Focus Question 1 can be examined. Once these methods have been illustrated, participants can work on the remaining questions in their small groups.

These questions explore the usefulness of breaking apart numbers and using the Distributive Property in multiplication. Focus Question 2 provides the opportunity to examine what happens when additively breaking both factors into two parts (a method frequently used to multiply two 2-digit numbers). In this case, the Distributive Property is applied repeatedly: $(a + b) \times (c + d) = (a + b) \times c + (a + b) \times d = a \times c + a \times d + b \times c + b \times d$. As participants examine a commonly made error in Focus Question 2, they will work on the connections between story contexts, the diagram solutions, the symbolic approach, and a multiplication algorithm. This error is exemplified by assuming that because $18 + 12$ can be determined by adding the tens $(10 + 10 = 20)$, then adding the ones $(8 + 2 = 10)$, and then adding to find the answer $(20 + 10 = 30)$, that something similar will happen when multiplying. However, 18×12 is *not* equal to $(10 \times 10) + (2 \times 8)$.

Focus Question 3 provides the opportunity for participants to examine closely the connection between a story context and a diagram as they work to explain a student's misrepresentation.

With Focus Question 4, participants consider what happens when the factors in a multiplication problem are additively broken in more than two parts.

Focus the whole-group discussion on Focus Question 2, "Why isn't the answer to 12×18 equal to $(10 \times 10) + (2 \times 8)$?" As participants show their methods of explaining this error and what would be correct, make connections between their various representations. For instance, ask participants how the elements of one participant's story context can be mapped to the diagram offered by another. In this discussion, participants can also examine the links between their representations and the procedures commonly taught for multiplying multidigit numbers and algebraic expressions. This discussion should provide strategies for participants to explore the questions concerning division in Part 2. See "Maxine's Journal" (pp. 186–193) for an example of such a discussion.

Break
(15 minutes)

DVD and Math activity: What happens with division?
(45 minutes)

Whole group (15 minutes)

Small groups (15 minutes)

Whole group (15 minutes)

This 5-minute DVD segment shows a small group of students explaining their solution to the problem: 198 students need to be seated in the cafeteria, and each table seats 14 students. How many tables are needed?

Before showing the DVD, give participants a few minutes to work on this problem and have them share their solutions with a neighbor. As participants are working on their diagram solutions, suggest they write out arithmetic statements that match their diagrams.

One solution might be

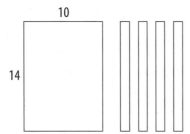

10

14

"First, I did 14 × 10 = 140. Then, I added tables of 14 until I had room for 196 students. I know I need 1 more table. The number of tables needed is 15. The solution can be thought of as (196 ÷ 14) = (140 ÷ 14) + (56 ÷ 14). That is 14 tables, and I need 1 more table for the other 2 children. 15 tables altogether."

Ask participants to view the DVD clip with these questions in mind:

■ How does the students' solution draw on the Distributive Property?
■ How do the teacher's questions make that connection more explicit?

Linking Addition and Division

Note that the Distributive Property holds in two directions.

$$(a + b) \times c = (a \times c) + (b \times c) \qquad (12 + 4) \times 2 = (12 \times 2) + (4 \times 2)$$

$$c \times (a + b) = (c \times a) + (c \times b) \qquad 12 \times (4 + 2) = (12 \times 4) + (12 \times 2)$$

One might notice that a similar pattern appears to result with division:

$$(a + b) \div c = (a \div c) + (b \div c) \qquad (12 + 4) \div 2 = (12 \div 2) + (4 \div 2)$$

However, in the other direction, the pattern no longer holds:

$$12 \div (4 + 2) \neq (12 \div 4) + (12 \div 2).$$

To give meaning to these symbols, it might be helpful to use a story context. For example, consider $(12 + 4) \div 2 = (12 \div 2) + (4 \div 2)$: "I have a class of 12 boys and 4 girls. I pair all students. How many pairs are there? The first expression on the left side of the equation finds the number of pairs by dividing the total number of children by 2. The second expression on the right side shows the number of pairs of boys added to the number of pairs of girls. Either calculation produces the same result: the number of pairs of children in the class."

Now consider a story context for $12 \div (4 + 2)$: I have 12 pencils to be shared among 4 girls and 2 boys. Each of the 6 children receives 2 pencils. However, the expression on the right side of the inequality—$(12 \div 4) + (12 \div 2)$—no longer represents the same context. That expression might indicate that 12 pencils are shared among 4 girls and another 12 pencils are shared between 2 boys.

While the symbolic expressions for these two division situations look similar on the surface, the first example can be rewritten as $(12 + 4) \times \frac{1}{2}$. Because this is a multiplication expression, the Distributive Property linking multiplication and addition can be applied. We have changed the domain of the numbers from whole numbers to rational numbers.

The second expression, rewritten as multiplication, would be $12 \times \frac{1}{(4 + 2)}$ or $12 + \frac{1}{6}$, which is not equal to $12 \times \frac{1}{4} + 12 \times \frac{1}{2}$. Participants are invited to explore these ideas in Problem 5 of the Math activity.

Participants' work with both multiplication and division continues with Focus Question 5 as they explore pairs of expressions to determine which are equivalent. While it is true that calculations alone will show which are equal and which are not, calculations do not illustrate the structure of the situations or answer why they do or do not work. Let participants know that story contexts are particularly useful for examining which pairs of expressions represent the same situations and which do not.

Focus the whole-group discussion on parts b and d in Focus Question 5. Once story contexts for these problems are detailed and participants have explained how part b represents equivalent expressions and part d does not, ask participants to state the general principle in this work.

One possible statement might be, "In a division problem, you can use addition to split the dividend and divide each part by the divisor and still maintain the integrity of the problem; however, if you split the divisor using addition, you change the meaning of the problem."

Problem 5 illustrates the importance of keeping symbols connected to what it is they represent. The pairs of expressions for parts b and d appear very much alike on the surface. It is when the meaning of the operations are considered that these differences become apparent. See pp. 194–196 for the discussion that took place in Maxine's seminar.

Math activity: Multiplying two negative numbers

(30 minutes)

Small groups (15 minutes)

Whole group (15 minutes)

Focus Question 6 presents an opportunity for participants to examine their thinking regarding multiplying negative numbers. Participants should begin by talking through the various strategies students used in Case 30. Encourage small groups to think over the formal argument reported by Ian and focus the whole-group discussion on his approach.

Begin the whole-group discussion by soliciting strategies that participants used to explore what happens when two negative numbers are multiplied. Include a strategy similar to that offered by Ian in Case 30 that is based on the Laws of Arithmetic and that illustrates that the rule, "A negative times a negative is a positive," has been adopted in order to maintain the consistency of the number system. See "Maxine's Journal" (pp. 196–198) for an example of such a discussion.

Homework and exit cards

Whole group

Distribute the "Seventh Homework" Sheet

Exit-card questions:

- What math ideas are on your mind at the moment?
- What was this session like for you as a learner?

Before the Next Session ...

In preparation for the next session, read the episodes that participants have written and write a response to each participant. See the section in "Maxine's Journal" on responding to the sixth homework. Make copies of both the papers and your response for your files before returning the work.

DVD Summary

Session 6: How many tables?

Fourth-grade class with teacher Jonathan (5 minutes)

Students are working in small groups on the problem: 198 children are to be seated in the cafeteria, and each table seats 14 students. How many tables are needed?

One group has drawn squares to represent each table. The students have drawn 7 squares along the top of the paper and another 7 squares in a row under those. They have placed the number 14 inside each square. Then they have one more square in a third row with a 2 in it. Jonathan, the teacher, asks them to explain the diagram and to make connections among the diagram solution, the arithmetic of their solution, and their final answer.

14	14	14	14	14	14	14
14	14	14	14	14	14	14
2						

Jonathan asks how students know they have 198 students represented. One student says she added 14 + 14 + 14 +...14. Another student says she figured

Multiplying in Clumps

out 10 × 14 first. Jonathan asks how students can see the 10 × 14 in the diagram and a student points to 10 of the 14 tables. Jonathan asks students to highlight that section. Then students write the arithmetic statements for 14 × 10 and 14 × 4 to show that 14 tables account for 196 students. They also write that there is a remainder of 2 students.

14	14	14	14	14	14	14
14	14	14	14	14	14	14

2

$$14 \times 10 = 140$$
$$14 \times 4 \ = \ \underline{56}$$
$$196$$

with a remainder of 2

Math activity and Focus Questions

Part 1: Exploring the Distributive Property

1. One way to calculate 28×6 is to decompose 28 into $25 + 3$ and then compute $(25 \times 6) + (3 \times 6)$. How do you know that 28×6 and $(25 \times 6) + (3 \times 6)$ produce the same answer? Use story contexts, diagrams, or array cards to illustrate this equivalence.

2. Use story contexts, diagrams, or array cards to explain why 12×18 is not equal to $(10 \times 10) + (2 \times 8)$. What adjustments are needed to make this correct? How do your representations illustrate the correct strategy?

3. In lines 473 through 477 of Case 29, Simone offers a story and a diagram to represent 19×4. Explain how Simone's representation does or does not match the arithmetic expression.

4. In lines 320 through 322 of Case 28, Miriam suggests that her students "draw a picture or use some other method" to see if Curtis's approach in lines 316 through 319 works for 24×25. Use story contexts and diagrams to explore this for yourself.

Part 2: What happens with division?

5. As you first look at the four pairs of expressions listed below, what similarities do you see? For each pair, do the two different expressions give the same answer? Use story contexts and/or diagrams to explore this question. Do any of these pairs lead to a generalization? How do you explain what is going on, both for those pairs that give the same answer and those that do not?

   a. $(12 + 4) \times 2$ $(12 \times 2) + (4 \times 2)$

   b. $(12 + 4) \div 2$ $(12 \div 2) + (4 \div 2)$

   c. $12 \times (4 + 2)$ $(12 \times 4) + (12 \times 2)$

   d. $12 \div (4 + 2)$ $(12 \div 4) + (12 \div 2)$

Part 3: Multiplying two negative numbers

6. In Case 30, Carl's students offer various explanations to explain why multiplying two negative numbers results in a positive product. Which of their explanations make sense to you? Why? What other explanations can you offer?

Optional Problem Sheet 3

These problems will provide you with the opportunity to experiment with symbolic notation. Feel free to turn in your work or questions on these problems at any time you would like comments or feedback. These problems will not be discussed during the regular seminar sessions.

1. For each of the following, do NOT solve for x. Instead, reason out if x must be zero, a positive number or a negative number. Use story contexts, diagrams, or representations, such as the number line and the charge model, to support your thinking.

   a. $3(x) + 5 = 21$ b. $3(^-x) + 5 = 21$ c. $3(x) + 5(x) = 21$ d. $3(^-x) + 5(x) = 21$

   e. $3(x) + 5 = ^-21$ f. $3(^-x) + 5 = ^-21$ g. $3x - 5 = 21$ h. $3x - 5 = ^-21$

   i. $3x + 5 = 6x + 5$ j. $3x - 5 = 6x + 5$ k. $3x + 5 = 6x - 5$

2. Consider each of the following statements. See if you can express, in words, the meaning of each statement. Then determine if it is true for all possible choices for a, b, and c. If the statement is always true, explain how you know. If it is not always true, explain why not. Use story contexts, diagrams, or representations such as the number line and the charge model to support your thinking.

   a. $a \times (b + c) = a \times b + a \times c$

   b. $(b + c) \times a = b \times a + c \times a$

   c. $a \div (b + c) = a \div b + a \div c$

   d. $(b + c) \div a = b \div a + c \div a$

   e. $(a \div b) \times n = (a \div n) \times b$

   f. $(a \times b) \div n = a \times (b \div n)$

   g. $(a \div b) \div n = (a \div n) \div b$

   h. $(a \div b) \div n = a \div (b \div n)$

   i. $(a \div b) \div n = a \div (b \times n)$

   j. $(a \times b) = (a \div n) \times (b \times n)$

   k. $(a \div b) = (a \times n) \div (b \times n)$

Seventh Homework

Reading assignment: Casebook Chapter 7

Read Chapter 7 in the Casebook, "Exploring Rules for Factors," including both the introductory text and Cases 31–34. As you read, note the generalizations that are present in students' work.

Write out the generalizations in words and/or symbols and bring them to the next session. These will be shared with other participants and also turned in to the seminar facilitator.

Writing assignment 1:

In previous seminar assignments, you have examined the mathematical thinking of your students or engaged with mathematics for yourself. At this point in the seminar, we are interested in how your ideas about teaching and learning mathematics have been influenced by the seminar experience. Please write about your ideas.

Writing assignment 2:

Look over the page titled "The Laws of Arithmetic." As we have stated in that document, we are interested in knowing your reaction to these formal statements and how they connect with the work we have done together in the seminar.

The Laws of Arithmetic[1]

About 100 years ago, after many centuries of humans thinking about numbers and figuring out how the operations work, mathematicians took on the project of codifying arithmetic in a way that is similar to Euclid's codification of geometry in ancient Greece. The goal was to determine a small set of axioms from which they could deduce the validity of all other claims. The mathematicians established nine laws that apply to the numbers on the number line, which can be used to justify any calculation strategy, as well as other generalizations. These laws are also used to justify all algebraic manipulations.

It is not appropriate to expect young children to participate in such a tradition. As we have been doing throughout this seminar, it is more likely and productive for students to justify the steps in their computations and strategies using basic models of the operations and corresponding representations.

However, as the seminar is drawing to a close, we are interested in how you connect with these laws. As you look at this list of laws, what do you make of them?

Commutative Law of Addition:

$a + b = b + a$ $\qquad\qquad$ $5 + 9 = 9 + 5$

Commutative Law of Multiplication:

$a \times b = b \times a$ $\qquad\qquad$ $3 \times 12 = 12 \times 3$

Associative Law of Addition:

$(a + b) + c = a + (b + c)$ $\qquad$ $(2 + 4) + 7 = 2 + (4 + 7)$

Associative Law of Multiplication:

$(a \times b) \times c = a \times (b \times c)$ $\qquad$ $(6 \times 2) \times 3 = 6 \times (2 \times 3)$

Distributive Law of Multiplication over Addition:

$(a + b) \times c = (a \times c) + (b \times c)$ $\qquad$ $(3 + 2) \times 4 = (3 \times 4) + (2 \times 4)$

Identity element for addition:

For every number a, $a + 0 = a$.

Identity element for multiplication:

For every number a, $a \times 1 = a$.

Inverse element for addition:

For every number a, there exists a number ^-a, such that $a + (^-a) = 0$.

Inverse element for multiplication:

For every number a other than 0, there exists a number $\frac{1}{a}$, such that $a \times (\frac{1}{a}) = 1$.

---

[1] This page is not meant as a full and complete codification of all arithmetic for the real number system. If you are interested in pursuing such an approach, we recommend *Contemporary Abstract Algebra*, by Joseph Gallian, Houghton Mifflin.

REASONING ALGEBRAICALLY ABOUT OPERATIONS

Exploring Rules About Factors

Mathematical themes:

■ What are the relationships between a number and its factors?

■ How can story contexts, diagrams, and cube structures be used to model relationships among factors?

■ How do we draw on such models to express and justify generalizations?

Session Agenda

Laws of Arithmetic	Whole group	30 minutes
Math activity and case discussion: Factors of factors	Groups of three Whole group	40 minutes 30 minutes
Break		15 minutes
Case discussion: Teachers' moves	Small groups Whole group	35 minutes 25 minutes
Homework and exit cards	Whole group	5 minutes

Background Preparation

Read

■ the Casebook: Chapters 7 and 8

■ "Maxine's Journal" for Session 7

■ the agenda for Session 7

■ the Casebook, Chapter 8: Sections 2 and 3

Work through

■ the Math activity and Focus Questions: Factors of factors (p. 246)

■ Focus Questions: Examining teachers' moves (p. 247)

Poster

■ Criteria for Representation-Based Proof

Materials

Duplicate

■ "Math activity and Focus Questions: Factors of factors" (p. 246)

■ "Focus Questions: Examining teachers' moves" (p. 247)

■ "Eighth Homework" (p. 248)

Obtain

■ index cards

■ cubes

Using the Term *Factor*

The term *factor* is used in two different ways in mathematical work. In computation, it is commonplace to identify either of the two numbers in a multiplication expression as a factor. For instance, in the equation $3 \times 5 = 15$, 3 and 5 are factors, and in the equation $\frac{1}{2} \times 12 = 6$, $\frac{1}{2}$ and 12 are factors. However, there is another usage of the term that is pertinent to this session. Within the field of mathematics known as number theory, which involves examining prime and nonprime numbers only positive integers are considered. A number is a factor of another number if it divides that number evenly without any remainder. In this use of the term, 3 is a factor of 15, 5 is a factor of 15, and 6 is a factor of 12, but $\frac{1}{2}$ is *not* a factor of 12. This is the meaning of the term *factor* as it is used in Session 7 of RAO.

Maxine's Journal

November 19

There were several things on my agenda for this session, and the first of my goals was to work some more on the idea of mathematical argument. This is a central idea that we have been examining, and one that is at the heart of mathematics. We have been working on it in the context of the resources available to elementary students—representations of operations. The generalization we worked on in this seventh session is that a factor of a number is also a factor of that number's multiples. Before we began, I hung the poster I had made up at the beginning of the seminar: "Criteria for Representation-Based Proof." I would refer back to it during our mathematics discussion.

Before we got to the work on factors, I wanted to hear about the participants' reactions to the list of the Laws of Arithmetic I had asked them to look at for homework.

Laws of Arithmetic

Throughout this seminar, we have been exploring generalizations that arise in the elementary and middle-school classroom, most of which relate to the properties (or laws) of the operations. In order to develop a deeper understanding of these properties, we have been examining them through various representations—cubes, area diagrams, number lines—and story contexts. These stories and representations allow teachers and their students to make meaning of these basic properties.

Once students enter a conventional algebra class, these properties are presented in their algebraic representations. The properties are what is given, assumed to be obvious, not needing exploration. Furthermore, these properties are among the nine laws that provide the bedrock for everything else in arithmetic and algebra. Every step of any calculation, any symbolic manipulation, can be justified on the basis of these laws.

Because these laws are so central to conventional algebra courses, I wanted participants in the seminar to see them through the lens of the work we have been doing. For this reason, as part of the sixth homework assignment, I asked participants to consider how these laws connect to the work of the seminar. Now, I wanted to hear what they had to say.

To help get their ideas flowing, I asked participants to talk in small groups, to share their thoughts and reactions to these laws. I was, however, most interested in the whole-group discussion, so I brought the group together after just 5 minutes and asked that they share their reactions. M'Leah said, "You know, as soon as I started to read these, I thought about how I learned the properties in high school. I was taught so many tricks, but I have a hard time making sense of them. Like, what does the Associative Law for Addition mean?"

I suggested we look at the Associative Law for Addition and asked M'Leah if she could help us begin thinking about the law. She said she has learned to pick some numbers, like 3, 4, and 7.

I wrote on the board, 3 + 4 + 7, (3 + 4) + 7, and 3 + (4 + 7), and suggested that participants make three towers of cubes, 3, 4, and 7 cubes high.

M'Leah said, "I don't get the parentheses."

I explained, "We can really add only two numbers at a time. So if you want to add 3 + 4 + 7, you can start out by putting the 4 with either the 3 or the 7. You have these three towers, and so when you combine them, you can start by putting the 4 on the 3 or the 7. One way, you get 7 + 7; the other way, you get 3 + 11. Either way, you end up with 14 because we didn't change the number of cubes."

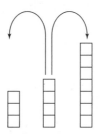

Jorge said, "But why bother making a law about that? Isn't it obvious?"

I said, "It's obvious in the same way that 3 + 4 = 4 + 3 is obvious. But let's consider subtraction. Does the Associative Property hold for subtraction?"

Together, we looked at (3 − 4) − 7 and 3 − (4 − 7). The two answers are ‾8 and 6, showing us that the law does not hold for subtraction. These numbers did give us some practice with new skills for working with negative numbers, but because participants' fluency with negative numbers is so new, these numbers still left participants feeling vague about the question. I decided to pick other numbers that would avoid negatives: 10 − 5 − 2.

$$(10 - 5) - 2 = 5 - 2 = 3$$
$$10 - (5 - 2) = 10 - 3 = 7$$

"Wow," Madelyn said. "I've never thought about that before."

June said, "But I'm still thinking about addition. When I see 3 + 4 + 7, I want to put the 3 and the 7 together to make 10, and then add the 4. That's what we want students to learn to do, too."

I told June that she is absolutely right. "For mathematicians, once you state these laws in this way, it's almost like a game to justify everything else using these laws. So, if you start with (3 + 4) + 7, but you want to add the 3 and 7 first, you have to go through a few steps. First, by commutativity, you can switch the 3 and 4 to get (4 + 3) + 7. Then you use associativity to get 4 + (3 + 7) and now can add 3 + 7 first.

"That's the game you play in formal mathematics. For young students, it doesn't make sense to try to make these distinctions. While we are working to make meaning for the operations of addition and understand how it works, students explore whether order matters, the way we saw in Chapter 3. If you have three numbers, you can add them in any order and get the same sum. Some students were saying that you could have more than three numbers. They said you can have a million numbers, and you can change the order without changing the sum. That's what we want young students to think about. Formalizing the system according to these laws can be worked on when they're older.

"I wanted to share this list with you more as a cultural experience for you. It's not something to take back to elementary students. However, as we've seen in the cases, some middle school students are ready to work with the laws."

I thought we were finished with this discussion, but then Claudette spoke up. "OK, I get all that. But I have another monkey wrench to throw in. I tried to understand the inverse element for multiplication. How does that rule work, if it does at all, with negative numbers? How can you multiply ⁻3 by anything and get 1?"

Claudette and Vera had both been absent for Session 6 when we worked on multiplying negative numbers. I decided it would be a good idea not just for them but also for the whole group to think about this again.

To start, I checked to see whether people were comfortable multiplying a positive number by a negative. They all agreed that 3 × ⁻1 = ⁻3.

I said, "Let's think about the number line for a moment. One way some people think about multiplying by ⁻1 is to flip the number line over. So if you're at 3 and multiply by ⁻1, you flip it over and land on ⁻3. If you're at ⁻3 and multiply by ⁻1, you flip it over and end up back at 3."

Denise said, "That's pretty neat." But several people in the group shook their head. That didn't work for them. Then I reminded the group of what Ian had done in Case 30. "We could think about it with the rules. What happens if you have (⁻3 × ⁻1) + (⁻3 × 1)?" I wrote that on the board. "You know that ⁻3 × 1 = ⁻3, right?"

Everyone agreed, so I went on. "You also know ($^-3 \times {}^-1$) + ($^-3 \times 1$) = $^-3 \times$ ($^-1 + 1$). That's the Distributive Property."

I paused to give participants a chance to look carefully at that equation. Once they understood its meaning, I finished the proof: ($^-3 \times {}^-1$) + ($^-3 \times 1$) = $^-3 \times ({}^-1 + 1$) = $^-3 \times 0 = 0$.

But, because $^-3 \times 1 = {}^-3$, we have

$$(^-3 \times {}^-1) + {}^-3 = 0$$

What can you add to $^-3$ to get 0? It has to be 3.

So $^-3 \times {}^-1 = 3$.

I concluded, "In order to keep the system consistent with these rules, it has to be that the product of two negative numbers is positive."

James said, "I still really like that." June added, "Now that I see this again, it's starting to make more sense to me. I'm writing this down and I'll try it with other numbers." A few others, too, conceded that seeing the argument a second time made it more convincing to them.

Going back to Claudette's original question, I concluded, "So now we can say $^-3 \times \frac{^-1}{3} = 1$; the multiplicative inverse of $^-3$ is $\frac{^-1}{3}$.

Vera was still shaking her head. She was still uncertain about multiplying negative numbers, so I suggested that we think about a context. "The tricky thing is, you need to figure out what negative means for both numbers."

Vera said, "Let's think about forgiving debts. That worked for me when we subtracted negative numbers."

We went with that, "So, let's say June owes \$3 each to Vera, Phuong, and Josephine, and all three forgive the debt. A debt of \$3 is forgiven 3 times."

Vera was smiling. "I like it," she said.

Kaneesha added, "I do, too."

Phuong explained, "Once those three debts are forgiven, you're \$9 richer than you were before." I pointed out that we had just talked about three different ways to make sense of why the product of two negative numbers is positive. "But this is a complicated idea. Remember, we spent lots of time thinking about subtracting negatives, and working out this problem will also require a good amount of time to fully understand. Multiplying negative numbers is something that would be good for you to spend some time thinking about on your own. We aren't going to do that today. We're going to spend the rest of today's session thinking about factors."

I had thought we were now ready to go on to the next activity, but James had another question, "It says here that these laws are used to justify all algebraic manipulation. How can that be? All the laws are about addition and multiplication. What about subtraction and division?"

"That's a good question," I said. "Do you recall that in Session 5 we were showing how subtracting is equivalent to adding the inverse?" I wrote, "8 − 3 = 8 + ⁻3" and "9 − (⁻4) = 9 + 4" on the board. Then I continued, "Once you're dealing with symbol manipulation, any subtraction can be changed to adding the additive inverse. Similarly, any division can be changed to multiplying by the multiplicative inverse. So, at that level, subtraction and division play a minor role, and mathematicians are basically dealing with just the two operations: addition and multiplication."

Now, we moved into the math activity and the cases participants had read for today.

Math activity and case discussion: Factors of factors

We say that one whole number, a, is a factor of another whole number, b, if a can evenly divide b. Another way to say it is that a is a factor of b if there is some whole number, c, such that $a \times c = b$. The word also has its own notation. To say 2 is a factor of 6, we write, "$2 \,|\, 6$." That is read as "2 divides 6," meaning that 2 can go evenly into 6. Although students in the cases did not use that notation, they were clear about what it means for one number to be a factor of another, and in Alice's and Jean's classrooms, students used manipulatives and diagrams to illustrate. Some of Jean's students also provided a context to help understand the concept. As a way to explain how they know that 25 is a factor of any multiple of 100, they say that any dollar amount can be changed into quarters.

However, when I taught this seminar before, some participants were unclear about what the word *factor* means and did not understand why you would be interested in it. They kept trying to turn the idea into a division problem in which you come out with a numerical answer. For this reason, I wrote the first problem in the Math activity about cases of soda. I wanted to provide a context in which you are not necessarily interested in the result of the division; you just want to know if the numbers divide evenly. If you have 384 bottles in cases of 24, can all the cases be full or do you necessarily have a portion of a case? I asked how the word *factor* relates to such questions. Before we got started, I told them that if they were unsure of that relationship after they answered the questions in Problem 1, they should call me over.

Small groups

When I stopped by Vera, Grace, and Yanique's table, Vera was saying, "24 is not a factor of 362 but *is* a factor of 384."

Grace added, "Because it's 16 whole cases."

Yanique nodded. The group was satisfied, so they moved on to the second question.

I saw that Josephine was working alone, while Risa and David were working together. I looked at Josephine's paper and saw that she was still working to figure out if 24 is a factor of 384. She had written:

☐   24

☐   24

☐   24

☐   <u>24   100 − 4</u>

☐   24

☐   24

☐   24

☐   <u>24   100 − 4</u>

☐   24

☐   24

☐   24

☐   <u>24   100 − 4</u>

☐   24

☐   24

☐   <u>24   75 − 3</u>

Josephine explained, pointing to the squares, "These are the cases of 24 bottles. I can think of four cases as being close to four 25s, which is 100. Each case is one bottle short of 25, so four cases is 100 − 4 bottles."

I stayed with Josephine for a while to talk through the fact that she now has 360 bottles; another case gives her 384; and so with 384 bottles you can have only full cases. However, if she has 362 bottles, there is necessarily a partially full case. I did not try to catch her up with Risa and David who were by now well into the third question. As I have seen throughout the seminar, Josephine is still working hard to develop flexibility in computation. Most participants in this group had achieved greater fluency through the first two "Number and Operations" modules ("Building a System of Tens" and "Making Meaning for Operations"), but Josephine must have started farther behind and needed more time to catch up. I know that she has learned a great deal in this seminar, even though she has not been able to take in some of our more abstract work. It is important to allow her to use the context of this seminar to move forward from where she is. At the same time, I must remain aware that she is not formulating generalizations about factors as she stays focused on strategies for adding 24s.

At the next table, Denise shook her head as she told me when she read the case title in Chapter 7, "Factors of factors," she immediately started picturing groups within groups. "But, when I worked on Question 2, which asked if all the factors of 48 are also factors of 192, I wrote out all the factors of 48 and tested them against 192; then, for the next set of numbers, I wrote out all the factors of 81 and tested them against 324. Why didn't I hold onto that picture that would have told me the answer from the beginning?" 220

I asked Denise what that picture was, and she told me it was having groups within groups. As she said it, she sort of moved her hands around to show what she meant, but it sounded too fuzzy for me to follow. I told her she was onto something, but I wanted her to be able to write it out, draw a picture, or show it with manipulatives—something that would be convincing to a skeptic. 225

I saw that M'Leah, Phuong, and Lourdes had written out the generalization stated in Alice's case—a factor of a factor of a number is also a factor of the number—and demonstrated it using arrays. Claudette, Madelyn, and Antonia were thinking in three dimensions, using cubes to illustrate their ideas. When this group looked back at the cases, they saw that the third graders had also used cubes, but they used them differently. Madelyn explained, "We were looking at dimensions of a rectangular prism as factors of the number of cubes. The students looked at the cubes and broke them into groups." 230

I asked, "What do you think about the work that Ben and Allan did with the cubes and Ben's early idea about solitaire? Had Ben come up with a proof at the very beginning of the case?" 235

Antonia said, "I don't think that connection to solitaire is really a proof. Solitaire doesn't say anything about factors, does it?"

I agreed with Antonia. It seems to me that the image of solitaire goes back to the notion that maybe a metaphor can help you think about an idea, but it does not offer a proof. To prove something about factors, you have to work with a model of multiplication and division. 240

Kaneesha, June, and Grace were struggling with their articulation of the generalization. "The thing is," June said, "it's really hard to come up with the words for the generalization." I suggested they look at the cases to see if they could identify any language that might help. 245

When Jorge, Mishal, and James were working on Question 3, I realized they had written out the generalization but had not developed a proof. "It works every time," is what they told me. I was disappointed that they were reverting back to the idea that coming up with lots of examples is enough to prove a generalization. Instead of pushing at this point, I suggested that they move on to question 4 and said, "I really want you to pay attention to what Ben and Allan did and see if that will lead to an argument in support of your generalization." 250

255

Whole group

When I brought the group together again, we first reviewed the terms, *factor* and *multiple*. Even though everyone had been working with these terms in small groups, it was still hard for many to specify what they meant. I explained

Exploring Rules About Factors

that the terms define "a relationship between two numbers." "You say that 24 is a factor of 384 because 24 divides 384 evenly, and 384 is a multiple of 24." I also clarified that when we use the terms *factor* and *multiple* in this way, we are thinking only about whole numbers. "Of course, $\frac{1}{2}$ goes into 4 eight times, but that doesn't mean $\frac{1}{2}$ is a factor of 4."

I went on. "Question 3 asks you to state the generalization implied by your findings in Question 2. Can you use these words, *factor* or *multiple*, to state the generalization?"

Instead of offering a generalization, Grace raised her hand and had me write out specific numbers:

Factors of 48

1×48	$\times 4 = 192$
2×24	$\times 4 = 192$
3×16	$\times 4 = 192$
4×12	$\times 4 = 192$
6×8	$\times 4 = 192$

She explained that this helped her see that all the factors of 48 are factors of 192. For any factor of 48, you can see what you multiply by to get 192. So, any of those factors of 48 evenly divides 192.

After discussing these numbers, we went back to the generalization. Charlotte said, "Well, actually it seemed like students in Alice's class had a way of saying it. 'A factor of a factor of a number is also a factor of the number.' At least, I think that's what they said."

Denise was not satisfied with that formulation. "It's confusing, the way you use *factor* so many times. I wrote, 'If n is a factor of m and m is a factor of c, then n is a factor of c.'"

Yanique said, "I wrote, 'If a is a factor of b, then all the factors of a are factors of b.'"

So far, we had three formulations on the board: Charlotte's, Denise's, and Yanique's. I asked whether all three said the same thing or if there are differences.

Antonia said, "It seems to me that Yanique's is more general than the others. She wrote about 'all factors,' but Denise wrote only that 'n is a factor of m.'"

I asked the group what they thought about Antonia's point, and several people argued in both directions. After a couple of minutes, it felt to me that further discussion would only confuse the issue, and so I said, "In fact, both Denise's and Yanique's statements say the same thing. When Denise writes 'if n is a factor of m,' it means that the statement is true for *all* factors of n."

Yanique said, "If n is a factor of m and c is a multiple of m, then n is a factor of c."

I asked, "What do you think of Yanique's formulation?"

Charlotte said, "That says the same thing, too."

Lorraine said, "If $a \times b = c$ and d is a multiple of c, then a and b are factors of d." [300]

At this point, I decided to introduce notation that was new to everyone. I wrote on the board "$x \mid y$." (I chose to use x and y so as not to get confused as the formulations on the board already used a, b, c, m, and n in different ways.) "This notation says x is a factor of y, or y is a multiple of x. You read it as 'x divides y,' meaning x goes evenly into y without a remainder. So, let's use that notation to write the generalization." [305]

It did not take us long to write the generalization, "If $x \mid y$ and $y \mid z$, then $x \mid z$."

On the board, we now had the following formulations of the same idea:

- A factor of a factor of a number is also a factor of the number.
- If n is a factor of m and m is a factor of c, then n is a factor of c. [310]
- If a is a factor of b, then all the factors of a are factors of b.
- If n is a factor of m and c is a multiple of m, then n is a factor of c.
- If $a \times b = c$ and d is a multiple of c, then a and b are factors of d.
- If $x \mid y$ and $y \mid z$, then $x \mid z$.

Charlotte commented. "The notation isn't scary if you know what it means." [315]

I then asked the group to share the arguments they came up with. Claudette spoke up first. "I wrote out 48 and 192 in terms of their prime factors: $2 \times 2 \times 2 \times 2 \times 3$ and $2 \times 2 \times 2 \times 2 \times 2 \times 2 \times 3$. Because 48 is a factor of 192, all the prime factors of 48 are contained in the prime factors of 192. Because the rest of the factors of 48 are products of prime factors, all those [320] will also be factors of 192."

Claudette's argument was different from what I had anticipated, but it was based on solid reasoning. I acknowledged that, and then I said, "You know, there are lots of things Claudette was relying on that we haven't discussed before. For example, she started out with the idea that any number can be [325] represented as a product of prime numbers. This is true, and it's important that, for any number, there is only one set of such prime numbers, regardless of how the factorization begins. She also pointed out that any other factor of 48 is a product of its prime factors. So, it's important to realize that Claudette's argument relies on all that machinery, and not everyone in the seminar has [330] thought about these ideas before."

I then turned back to Claudette, "You explained how you know that all the factors of 48 are factors of 192. Is that the extent of what you were trying to show?"

Claudette responded, "No. Even though I talked about 48 and 192, you could do the same thing for any other number and a multiple of that number. [335] I could say y is a multiple of x, and then say the same things about x and y. It just seemed easier to see with 48 and 192."

I nodded and said, "We have seen things like that before. A couple of sessions ago, we talked about Ian's argument for why $^{-}3 \times {}^{-}4$ must be equal to 12 but thought of that as an argument for why the product of any two negative numbers must be positive. Some people call this an 'algebraic use of numbers.' It is an argument that could be generalized.

"But as I said before, Claudette's argument relies on lots of machinery that isn't necessarily present for everyone in the room and most likely isn't available to most elementary students. So let's look at some of the other arguments you came up with."

Madelyn then presented her group's way of thinking about it. "I thought of a rectangular prism with 192 cubes, 48 in each layer. Because there are 4 layers of 48 to make 192, I know that 192 is a multiple of 48. So, let's say I arranged the layers so that they're each 6 × 8. Because 6 is the length of one dimension of the prism, 6 must be a factor of 192. But you can do that with any factor of 48."

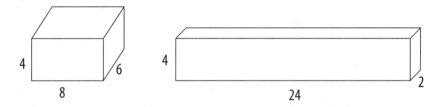

Madelyn had made the case that all factors of 48 are factors of 192, but she had not yet extended the generalization. Before I could make that point, M'Leah wanted to share her method, so I went on with the group and would get back to that point later.

M'Leah said she thought her way was like Madelyn's, but it was in two dimensions. "You start with a 48 × 4 array. The area of the rectangle represents a number, and the sides are factors of that number. So, the area is 192 and 48 and 4 are factors. Then, because 24 is a factor of 48, you can cut the array into two pieces, 24 long, and move half of it down to make another array. The array still has 192 squares, but the sides are now 24 and 8. So, 24 and 8 are also factors of 192."

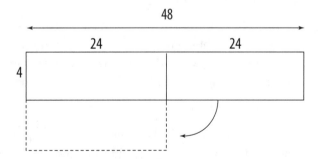

M'Leah went on, "You can do that with any factor of 48. Like, here I showed it with 24. But let's look at 6. If you start with a 4 by 48 array, then you can make lengths of 6 along the top, and then bring all those pieces down.

You have another array whose area hasn't changed, but the dimensions are now 6 by 32."

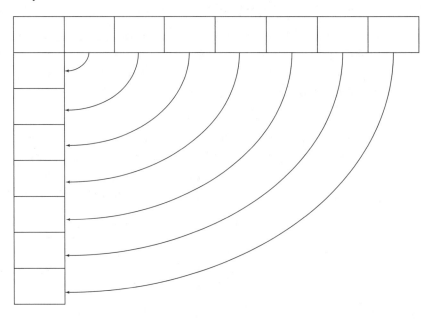

At this point I asked participants to look at the poster, "Criteria for Representation-Based Proof." I wanted them to consider Madelyn's and M'Leah's arguments with regard to these criteria.

1) The meaning of the operation(s) involved is represented in diagrams, manipulatives, or story contexts. 2) Representations accommodate a class of instances (for example, all whole numbers). 3) The conclusion follows from the structure of the representation. I said, "Let's first look at what Madelyn did."

Madelyn spoke up for herself, "Factors and multiples are about multiplication. My representation shows multiplication with an array. If you have a number represented in an array, each side is a factor. If you have several layers of that array, you have a prism and that prism is a multiple. So, because 6 is a side of the layer, 6 is a dimension of the prism and the prism is $6 \times 8 \times 4$. So, 6 is a factor of 192."

I said, "OK, so we understand how your representation satisfies the first criterion. What about the second?"

Madelyn continued, "Like Claudette said, it's easier to talk with particular numbers. But those could have been any numbers. You can start with the bottom array being any number, and it can be arranged with any factor as a side. You can take any multiple of that array to make the prism."

I asked, "And the third criterion?"

Madelyn finished, "We said one side represents a factor of the number arranged in an array. The multiple is the whole prism, multiple layers of that array. If you take the dimension of that array and multiply it by the other two dimensions of the prism, you get the whole prism. So, the conclusion follows from the structure of the 3-D array."

As a group, we worked through M'Leah's argument in the same way. Then Risa said, "I thought of it differently. I just thought, 6s are in 48 and 48s are in 192, so 6s are in 192."

I was concerned about what Risa considered to be proof. Although she referred to numbers, her words sounded to me like the metaphor, "Flour is in bread and bread is in stuffing, so flour is in stuffing." (In a previous seminar, this metaphor had been offered as justification for the generalization we were now discussing.). I didn't hear anything about multiplication in Risa's argument. Before I spoke, I saw that Grace had a question. "Did you come up with a picture or anything? I don't really get what you mean."

Risa said, "I didn't have a picture in my head. That's just what my thought was."

Denise said, "That's the way I was thinking about it before, but Maxine told me I needed to convince a skeptic. Just saying that won't convince anyone who doesn't already believe it."

Risa said, "OK, so I don't need a picture, but I can say there are 6 things in each bag, and a bunch of bags make up 48. So I've got one bunch of bags, and I can take another bunch and another bunch and another bunch. I've still got a slew of bags with 6 things in each bag. How's that?"

I paraphrased Risa's description of an image for the group, showing with my hands the bunch of bags and moving a step over for each additional bunch of bags. Then I asked Risa's question, "How's that?"

Charlotte said, "That's a neat way to think about it, too. However many are in your bunch, you take that many bags each time to make your multiple."

We again went back to the criteria: 1) Multiplication was represented by copies of equal groups, either multiple copies of a bag each with 6 objects or multiple copies of bunches of bags. 2) The bags might have any number of objects in them; you can have any number of bags, and you can have any number of bunches of bags. 3) Because any number of copies of bunches of bags is made up of some number of bags, the conclusion follows from the structure of the representation.

Kaneesha raised her hand and declared, "I just have to tell you of my new discovery. When you just went through Risa's way of thinking about it, I realized that the number of times it takes to make your multiple, you use that same number times what you originally multiplied your factor by. Do you know what I mean?"

Actually, I did know what Kaneesha meant, though it is hard to keep track of what to call all those numbers. I asked if anyone else followed what Kaneesha was saying.

James said, "Yeah. It's like you have $8 \times 6 = 48$ and $4 \times 48 = 192$. You take your 8 and you multiply it by that 4, and that's now what you have to multiply by 6 in order to get 192. $32 \times 6 = 192$."

I nodded and suggested we look at the symbols a slightly different way. Kaneesha's observation is that $(4 \times 8) \times 6 = 192$, and that was because $4 \times 48 = 192$, or $4 \times (8 \times 6) = 192$. So $(4 \times 8) \times 6 = 4 \times (8 \times 6)$. "Does that look familiar?"

Denise nodded, "Yep, that's the Associative Property again."

Phuong added, "Isn't that what Tyrone was saying in Jean's case? Look at line 494."

Vera said, "I couldn't follow Tyrone. Is that what he was saying?"

I suggested that participants might want to take another look at what Tyrone said and how Jean, the teacher who wrote the case, thought about it in terms of algebraic notation.

Mishal led the way. She came up to the board and wrote out the algebraic notation as she explained, "If you look at line 504, it's as if a is the number of things you have in a bag; b is the number of things you have in a bunch of bags; and c is the number of things you have with several bunches of bags. At line 505, it says that m bags make up a bunch; that's

$$m \times a = b.$$

"Then you have n bunches; that's

$$n \times b = n \times (m \times a) = c.$$

So that means, you have $n \times m$ bags, with a things in each bag, in those several bunches;

$$(n \times m) \times a = c.$$

Phuong still wanted to show her proof. She said, "I have another way to show it without having to think about any particular numbers. Let's say a number is represented by this rectangle." She held up a rectangle she had cut from paper. "If another rectangle is a factor of the larger number, then I can keep folding over that rectangle and it will come out without a remainder." Then she showed us the folded paper.

Phuong went on with her explanation. "Then, if an even smaller rectangle is a factor of the middle rectangle, I can keep folding it over, and it will come out without a remainder." She folded the paper to show us the smaller piece.

Exploring Rules About Factors

Phuong then opened up the paper so we could see the folds. "And this shows how the smallest rectangle is a factor of the big rectangle." 465

Although they do not think of using paper folding the way Phuong does, the other participants always appreciate her demonstrations. I heard "Aaah" and "Wow, that is so cool."

Grace said, "Wait a minute. I'm not sure I get it. What is the big rectangle?"

Phuong held up her partially folded paper. "This rectangle is the number we start with, y, and where you see the folds here, that shows the factor of the number, x." 470

Then she unfolded the paper again to show the long rectangle. "The long rectangle is a multiple of the number; that's z. You can see that the factor of y is also a factor of z." 475

Kaneesha said, "I still like Risa's way. You have a bunch of bags with x things in them and the bunch altogether is y things. Then if you take a bunch of bags and another bunch of bags—it doesn't matter how many bunches— you end up with a multiple of y, and that's z. But it's still a slew of bags with x things in each, so x is a factor of z." 480

Before going on, I said, "We've now looked at several different arguments that all support our generalization, if $x \mid y$ and $y \mid z$, then $x \mid z$. Take a moment to look over all these different ways of thinking about it. How do you feel about the different arguments?"

Claudette immediately spoke up, "Thinking about prime factors was my way into the problem. But when I look at all of these different ways, they all make sense to me and make me feel more solid with the ideas. Now the generalization isn't just about prime factors, but it's also about prisms, and cutting up arrays, and bags of goodies. And I can look at the end of Case 33 where it's written out in algebraic notation, and it makes sense, too." 485

490

M'Leah said, "When I first saw what Madelyn did, I thought her way showed it only for the numbers 6, 48, and 192, but my way showed the generalization. It was really important for me to see how Madelyn's way can also generalize."

Charlotte said, "I'm sitting here soaking up all these different ways of proving the same thing. But the point that's striking me most right now is that having seen all these visual representations the algebraic notation is making sense."

I nodded and said, "It's not just looking at all these different arguments, each in isolation, but looking at them together. If we look across representations, each can give the other more meaning."

I suggested we spend a moment considering the students in the cases. "So, let's think about what those third and fourth graders were doing. Did they prove a generalization?"

M'Leah said, "I think the third graders had the same idea I did. They made an array with cubes. So they had a bunch of sticks of cubes, and one stick was a factor of the whole array. Well, if you can break up one stick into the same size group, then you can break all the sticks into that size group—and that shows that size is a factor of the whole thing."

Charlotte said, "I'm not sure all students saw the generalization."

M'Leah said, "But look at line 294. The teacher said that many of the students understood what Allan was saying. Look at what Karen says in line 311: 'You can do it with every number.'"

Charlotte said, "Maybe it's a generalization for Karen and Allan and Ben and a few others. I don't think we can say that about the rest."

Denise said, "So here's my question again. You've got a few students doing this wonderful stuff. What about students who aren't getting it?"

I said, "I think this is a good question, and I'd like us all to consider that. Let's assume (although we don't know) that only a few students in Alice's class were able to think in terms of a generalization. What might be happening with the rest of the students?"

Charlotte said, "Alice writes that all of this was happening when students were working on factors. Everyone in the class had lots of opportunities to practice finding factors."

Kaneesha said, "You know, for students who are having the most difficulty, they're working with cubes and developing those images of what it means to be a factor. Even multiplication is pretty new to these students, and they're getting lots of practice thinking about multiplication."

Vera added, "And I bet it's important for them to see Ben and Allan's argument. Even if they can't think in terms of generalizations, they're looking at those cubes and seeing how you can reason with them. Maybe some students are thinking only about those particular numbers: that the factors of 8 are also factors of 120. But they could be getting a lot from these explorations too. They're just not ready to generalize yet."

Exploring Rules About Factors

James commented, "The thing is, you need to be careful in making up small groups. What if someone who was slower was put with Ben and Allan? Ben and Allan might be racing along, and that student wouldn't have much opportunity to learn. If there was another student who just really needed practice finding factors, they would be good together."

These were all good points and were moving us into the next discussion about teachers' moves. So I suggested that the group take a quick break, get their snack, and sit down again with their small group.

Case discussion: Teachers' moves

I distributed the activity sheet, "Examining Teachers' Moves," and said, "I'd like you to take another look at the cases, this time with an eye toward the teachers' moves." In this activity, I wanted the small groups to think about how, in each of the cases, the teacher's agenda and her attention to her students' thinking drove her actions. I listened in on small groups and saw that most of them were discussing what took place in Alice's two cases. Each group identified different moments, but I thought it would be helpful if we could look at several of these, to see what was behind all of them. So that is how I opened the whole-group discussion. "Let's start out by listing some of the moves Alice made that were of particular interest to you." In order to keep the discussion focused on specifics of Alice's moves, I added, "Remember to identify the line number so we can all look at the text together." Denise started us off. "I was interested in how she started this whole thing at line 34. She posted three statements: All of the factors of 100 are also factors of 300; most of the factors of 100 are also factors of 300; and some of the factors of 100 are also factors of 300. Then she wanted students to say which of these they agreed with and why."

Charlotte said, "I'm actually interested in something that comes before that. In the first few paragraphs of the case, starting at line 2, Alice discusses this idea that has been bubbling up in her class for some time. I think we should pay attention to that as a teacher's move—paying close attention to what students are saying and recognizing the common thread. She sees that all of these different observations her students make are about factors of factors. So, the move Denise pointed out stemmed from what Alice has noticed about her students' ideas."

I felt grateful to Charlotte for bringing this up. She was focusing the group on the teacher's agenda and the teacher's attentiveness to her students' ideas. Without her, I would have needed to draw attention to these same points.

Madelyn added, "I agree with Charlotte. What Alice says in the beginning is very important, and I'm interested in the move Denise pointed out because it allows Alice to see where different students are with the idea; which of her students are ready to make claims about *all* the factors of 100. She's about to go after a much bigger generalization, but she starts out seeing where they are on this narrower one."

I mentioned that this idea of "Is it always true?" still seems to be daunting to some of her students.

Jorge said, "I was looking around line 59. Alice says that it's time for the class to start exploring the generalization, so she comes up with a number, 48, that has interesting qualities. It will challenge them. This number lends itself to this exploration."

M'Leah added, "She has picked a manageable number that everyone can work with."

Lorraine said, "We need to look at line 55. She has stated the generalization she is after. She said, 'If n is a factor of m, then all the factors of n are also factors of m.' That's what's guiding her."

Lorraine was highlighting the generalization that was driving Alice throughout the two cases. I added, "We have experienced how challenging it is to articulate some of these generalizations for ourselves. It's important that Alice was able to be so clear about the idea she wanted her class to think about."

M'Leah said, "So then she got them started with that question but focused on the number 48. She started by giving them more structure. Each pair picked a factor of 48, and then she asked, 'Is it true that all the factors of ___ are also factors of 48?'"

Kaneesha said, "I think she's very conscious of what she's doing. She has a plan. Students might move to the left for a bit, or to the right, but they are still on the road. Like at line 89, students didn't get enough cubes, so she made a decision. She might have just let them get started and then discover they need more cubes. She might have told them they need 48 cubes. Instead, she asked them to think about how many cubes they needed. And I want to jump ahead to around line 222. In the class discussion, Karen asked about 0. Alice noticed that—she said questions about 0 often come up—but she decided to leave that aside for now. She sees that getting into 0 now would take them off the road. So she makes the decision not to address it at that time."

Lorraine continued to think about the generalization that was guiding Alice. "Throughout the case, she keeps asking the class, 'Is this something special about some numbers or is it true about any number and its factors?' She asked that at the end of the whole-group discussion, around line 160 in the Casebook, and she asked it at the start of the next day around line 167."

June said, "I was interested that she doesn't seem pressured to get it all done that same day. She continued with the ideas over several days. Even at the end of the second case, it sounds like she'll come back to this later."

Denise was also struck by this feature of Alice's teaching. "Around line 296, she says, 'Let's let that sit with us all for a few seconds.' She's not rushed to

Exploring Rules About Factors

get somewhere, even though it's clear she has a direction and a goal. She gives her students time to think." 615

Charlotte added, "She chooses particular moments when 'think time' is especially fruitful. That's also a way of alerting students that something important has been said."

Grace brought up a new point. "I'm interested that Alice recognized a way of thinking in students' work about factors of factors that could lead to a proof. It started with Fran and Holly, and then at line 134, she said that Susan and Marina built on that same idea. Ben and Allan also used the same idea." 620

Madelyn said, "But, you know, some students were thinking they should keep testing numbers. On line 199, Alice asked how many numbers they'd have to try before they were convinced the rule always worked, and Susan said one or two more. But after a while, she reminded the class of the work they did with switch-arounds. Then they used models to convince themselves that switch-arounds for multiplication always works." 625

630

Yanique said, "You know, I teach third grade, too. From the case, it's clear that a lot of students were working on this generalization, and some even came up with a good proof. But I'm pretty sure that not all third graders in a class would be there. The thing is, even if they weren't working on proving the generalization, they would still be doing good work for third graders. Maybe some students couldn't hold all those ideas together; they were still working on the factors of 48 or the factors of 120, and doing all this multiplication. I'm back to the idea that Jorge and M'Leah were talking about in the beginning of this discussion. When Alice started the investigation, she chose the number 48 because it has lots of factors to work with, but it is small enough for everyone to get into. I bet for all those sessions, even if some students couldn't follow everything that was going on, everyone had a chance to work on what they needed." 635

640

I was impressed by this discussion. I had been expecting to do a lot of work to draw the group's focus to Alice's overall agenda and her attention to her students' thinking, but some participants—notably Charlotte and Lorraine— did some of that work for me. Their observations about Alice's teaching moves were significant, and now I was interested in how they thought about implications for their own teaching. However, I didn't have time to ask. I will have a chance to read what they wrote on this topic for homework before the next session. 645

650

Exit cards

For tonight's exit-card activity, I asked people to write about something from the session that struck them. Predictably, the responses were quite varied.

Claudette wrote, "The one thing that stood out for me is while doing the math activity trying to see if 81 was a factor of 324, could all the factors of 81 also be factors of 324? I made a tree and finally understood the real purpose of it. If the bottom floor fits to the second floor, all could fit under the top floor.

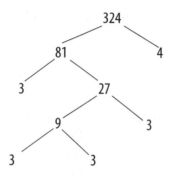

"I remember using this to find prime factors. It all made sense now."

Lorraine wrote, "I found it amazing to read that these nine Laws of Arithmetic are the basis for any calculation strategy. Sometimes what we think we understand can be difficult to put into words. In certain cases, the algebraic notation is clearer to me."

Denise wrote, "The cases absolutely amazed me. There were so many things in place that made these math classes really work. (I can't help but think how I would have loved to be a student in such a class—but that type of learning would not have happened when I was a kid!) So what makes it work now and why in some places but not in others? What do teachers (me!) need to know to bring kids to such rich exploration of ideas? I really like looking at the teacher's moves. Doing so has become something that I pay close attention to when I am reading because I want to be able to teach like this. I want my kids to be able to learn like this. I also liked when a child said, 'I revised.' She had encountered new evidence and was free to change her mind. Not, 'I was wrong,' but simply, 'I see this differently now.' What power of thinking for students to feel that there isn't a fixed end point but always the option to learn more and develop ideas further. Again, incredible cases—so much to learn."

Phuong wrote, "The discussion about teachers' moves could be very rich. At our table, people seemed to be so tired and distracted that it was hard for us to stay on track and to go deeper into the analysis of numbers, factors, multiples, images, and who said what. I was glad we had the whole-group discussion because that way I could hear from people who really did think about it, but we didn't do good work at our table."

Exploring Rules About Factors

Responding to the seventh homework

November 24

When I gave out the "Seventh Homework" Sheet, I recognized that we were approaching the end of the seminar, and I wanted participants to start thinking more generally about the lessons they were learning. I asked that they consider how their thinking about teaching and learning mathematics has been influenced by the seminar. Also, I distributed the sheet that listed the Laws of Arithmetic and asked that they respond in writing to those. Now, having read what they wrote, I find myself thinking back to my first impressions of these participants. For example, in the first session, Risa was so ready to jump into algebraic notation and needed encouragement to examine a variety of representations. Also, I think about her first student-thinking assignment in which she asked so many rapid-fire questions of her students, she didn't allow herself time to find out what they were thinking. Throughout these past seven sessions, I could see changes in her approach. From what she wrote in this assignment, I know that Risa can see them, too.

Risa

Writing Assignment #1

How my ideas about teaching and learning mathematics have been influenced by this seminar's experience.

I have always tried to "see" more rules and numbers when teaching and learning about math. Over the course of my 36 years of teaching, I have participated in more math-related workshops than in any other type. Often, however, I have been disillusioned by the lack of an in-depth look at a particular rule or mathematical idea. After our first meeting, I wasn't quite sure if I wanted to continue (too far . . . too late . . . too tired . . . too old . . . I'm thinking of retiring, and so on). I'm glad I stuck it out.

    Algebra was one of those subjects I didn't enjoy exploring in high school. The teacher was a cranky young nun who didn't like everyone in the class. She had her favorite students who were always called upon to plug in those numbers while the rest of us were referred to as "You goose!" with such contempt. I did well because I paid attention and learned how to fill in the blanks, never thinking about why all of the numbers seem to fit into place.

    Over the years, I've used x's and y's in number sentences, talked about the Commutative, Associative, and Distributive Laws and the Properties of 0 and 1 without ever considering the thinking of my students. I guess I didn't even realize that I didn't fully understand the concepts myself. So, I never thought about how to present the ideas to my students. Never had algebra been talked about in any of the classes I took during my teacher training. I still used a book that I have from my math methods course each time I had to teach "algebra rules." I reviewed, memorized, and delivered, expecting students to remember.

Did they really understand? I never knew. I thought that if they "plugged in the numbers," they got it. Well, I finally got it when I began to talk to my students about their ideas and asked them to convince me about what they were thinking. I was often amazed at how much they were able to think about an idea if I gave them the opportunity. Now, when we talk about relationships and generalizations, it just feels natural to dig deeper and allow for discussions about another way to "look" at what's happening to our numbers.

Writing Assignment #2

The Laws of Arithmetic

Does the work we have done in this seminar allow me to bring any (new) meaning to these laws? They finally make some sense! Now, I can picture some of these laws in my mind. When I think about the Commutative Law of Multiplication, I see arrays and factor pairs. I can easily manipulate the numbers in my mind. I get so excited when my students and I explore multiplication, and they are able to chunk numbers or break them down with ease. I am proud that they are able to find the solution or find the part that's missing when they use the Distributive Law of Multiplication. They all don't get it yet, but boy, those who do have gained so much power, and they have the confidence to tell me more about their thinking. So it's not just about what I have learned and discovered, but what my students have as well. I can now look at most of these laws and feel comfortable talking about the ideas they contain, asking for representative models, and expecting students to begin to have a discussion about them.

I guess I don't need to refer to my college book every time I want students to think about the laws and properties. I feel comfortable exploring and figuring out why $a + b = b + a$.

In my response, to Risa, I wanted to affirm the shift she has described.

Dear Risa,

It's great that you are feeling so comfortable about the Laws of Arithmetic. I am so struck by how central "seeing" is to your understanding. You often use the words "look" and "see" in your writing. It seems to me that developing mental images of the operations is very important to you and that you have come to understand the importance of this for your students as well. I think that the connections you are helping your students make to seeing and modeling how and why these laws work are so critical to their understanding. It is also heartening to hear that you can see the connection between students' work with properties (or laws) and the power that gives them.

I have really appreciated your energy and reflective participation in the seminar. It sounds like you are having—and will continue to have—very fruitful discussions with your students. I am glad you stuck it out, too!

Maxine

While Risa's writing was full of what she had learned and conveyed a sense of satisfaction, Phuong wrote about the many questions she still has. Some of those come from her work as a math coach.

Phuong

How my teaching and learning mathematics have been influenced by the seminar experience.

■ *Many of the mathematical concepts may seem obvious to me at the first glance, but it has always been helpful for me to see how concepts are understood, misunderstood, explained, and represented in so many different ways, depending on the entry point and learning style of each participant. So, in this sense, it has helped me to deepen my understanding in mathematics. For example, during the discussion of negative numbers, it was easier for me to explain why $^-3 - 7$ is $^-10$ using a story such as if you owe \$3 and then you owe \$7, altogether you owe \$10. But Maxine's idea about the relationship between the negative sign and the direction along the number line is different, and I have not thought of explaining this concept in this way.*

■ *Some of the cases have helped me to appreciate how difficult the job as a teacher could be, while in the midst of a class, when there are ten different directions one can go at every moment. The importance of understanding students, the purposes of the lesson, the current big ideas/overarching ideas, and the direction/next step the whole class/individual students are going are all important deciding factors that help to shape the progression of a lesson. It is indeed a very complicated process.*

■ *It has also been an interesting experience for me to see how teachers learn to rethink algebra. Sometimes I do take it for granted that many of these concepts seem to be pretty straightforward to me. It is helpful for me to watch how some concepts could be slippery and confusing for adults.*

■ *I began the seminar with a goal of learning from Maxine how she teaches algebra. I have many questions about learning and teaching mathematics. I often struggle with constructivism and wonder if this is a theory for teaching or learning or both. When is telling appropriate, what is too much, or not enough? What do we need to give to learners so that there is enough for them to build upon and to make sense out of? When does "giving too much" become an impediment to the process of expressing and exploring concepts on the part of the learners? Are there any guidelines to these questions, or does giving any guidelines violate the very premises of the pedagogy? This is some of the struggle that I see myself going through every day. In Case 26, line 122, the teacher said, "Because time was running out, I decided to share another answer." Does it mean that if there were enough time, the teacher would not have put out the answer? Is the teacher being apologetic for giving out too much and use the lack of time as an excuse to give/suggest the answer? (See, I am not the only one to use the excuse of not having enough time!!)*

- *Are modeling and demonstration necessarily bad ways to teach and imitating and copying bad ways to learn? In reflecting on my own learning in mathematics, I have found that if I am unable to digest the material and make it my own, learning does not take place. There is no transfer, generalization, or creation of new layers of meaning and knowledge. Through exploring things myself, I enjoy the learning process and take pride in what I have discovered. On the other hand, I also learn a lot by watching and imitating what other people do, with understanding. All these processes are just as valuable to me. I often ponder what "standing on the shoulders of giants" means. When is "standing on the giant" enabling, and when does it become disabling?*

- *In class, as I watch and listen to Maxine, I catch myself thinking, "Ah, so even Maxine gives this away!! Ahh, what a beautiful move! A great leading question!" Teachers are responding to it and coming up with a lot of their own ideas.*

The Laws of Arithmetic: Does the work we have done in the seminar bring any new meaning to these laws?

I have been using these basic arithmetic laws without paying conscious attention to what they really mean. This particular seminar required me to slow down and look at these laws from a different perspective.

1. *Commutative Rule: It was neat to take a fresh look at it and ask myself, is 5 + 5 + 5 + 5 + 5 + 5 + 5 the same as 7 + 7 + 7 + 7 + 7? Seven 5s and five 7s—quite beautiful to be confident that they are equal.*

2. *The Commutative Rule for both addition and multiplication once again helped me to see how moving the numbers around can sometimes make the computation a lot easier.*

3. *Distributive Property of Multiplication: Again, it was beautiful to see how the use of visual image of the array can help to dispel so much misunderstanding. It is helpful also to be able to work with this concept from either direction—solving a word problem based on the equation, and asking participants to create a story to go along with an anonymous multiplication equation.*

4. *What I don't understand is why we spent so much time discussing the identity elements for addition and multiplication. I might be missing out on something that is significant, and yet I am not getting it.*

Whew! Phuong packed a lot into this assignment. I had realized before that she did have lots of questions, and I'm glad she took this opportunity to articulate some of them. However, I could not address in one response everything she wrote about, so I chose to make a few points.

Dear Phuong,

You say, "I often struggle with constructivism and wonder if this is a theory for teaching or learning or both." I think most people will agree that it is a theory of learning (though they don't necessarily agree on what that theory is), and although some people talk about "constructivist teaching," others vehemently argue that it is not a theory of teaching. Over the last decade, "constructivism" has become an overly used word with a high degree of charge. In fact, over the last few years, I try not to use it at all. 850

Instead, it seems to me to be more productive to dig underneath the idea, to try to be clear about what we mean by "understanding mathematics," and to try to figure out how students will achieve that understanding. One thing 855 *that is clear is that it cannot be assumed that a student will learn something simply by listening to what the teacher says. Sometimes, when a teacher explains something, students will learn—but you can't assume that will always happen. Most teaching in past decades has had very modest goals for student understanding and has relied almost exclusively on the teacher showing* 860 *students what to do and students practicing that method As we take on more ambitious goals for understanding mathematics, it becomes clear that this view of teaching is insufficient. The thing is, other simplistic stances such as giving students problems without telling them how to solve those problems or having each student present a strategy to the class are as likely to be unsuccessful.* 865 *Teaching for understanding is a complicated process.*

How do we help teachers become better at teaching for understanding? One argument I often make at professional meetings with other people who design teaching materials is that, initially, it is important to set aside discussions of pedagogy. The tendency, within the current cultural milieu 870 *of teaching, is to judge teaching moves independently of the mathematics at hand or of what students are learning. That only encourages such simplistic rules as "don't tell." Instead, I want teachers to learn to look at a classroom event and first identify the mathematics at issue and attend to students' mathematical ideas. Only after teachers are able to do that does it seem* 875 *fruitful to look at teachers' moves. Now, they can consider teachers' moves in relation to the mathematical agenda and student learning.*

You point out that, in Case 26, line 122, the teacher says, "Because time was running out, I decided to share another answer." We don't know exactly what she was thinking and whether she was "apologizing," or whether she 880 *wouldn't have put out her own answer anyway. Perhaps that is beside the point. We can still ask, what did students do in response to her move? Did she introduce an idea they could connect with? What was that idea? How might she check in with her students later about that idea?*

Certainly, one factor to be considered in any teaching event is how much 885 *time is available. Long periods of time allow for more extended investigations that give opportunities for teachers to learn more about how students are thinking. When only a short amount of time is available, there may not be an opportunity to learn how students are thinking, and so a teacher must make her move and hope for the best. It seems that's the situation you're often in* 890 *when you are in your coaching role. If you have only a few minutes to meet with a teacher, how can you best spend that time? Is it best to pose a question for the teacher to think about? Or is it better to simply give the teacher information? Or should you point out something you noticed among her students or in her teaching? Something that might interest her about the curriculum? Perhaps if you see a teacher repeatedly, even if only for a few minutes each visit, you will learn enough about her that you can make more informed decisions.* $\quad$ 895

Phuong, it is clear that you have a good understanding of mathematics, and you enjoy making new connections. The images you come up with for illustrating mathematical ideas are very insightful. From reading your cases over the last two months, I have just one suggestion. When you work with a small group of students, it seems that you are able to move forward with the stronger students, but the weakest student is left behind. My suggestion is that you attend to the student you perceive to be weakest and find that student's mathematical strength. Once you find what that student does understand, determine what steps you can make with him/her in the direction of your mathematical agenda. $\quad$ 900 $\quad$ 905

It has been a pleasure to work with you in the seminar.

Maxine $\quad$ 910

Detailed Agenda

Laws of Arithmetic   (30 minutes)

Whole group

Give participants 5 minutes to talk to a neighbor about their reaction to the Laws of Arithmetic. Participants are apt to have mixed reactions. Some might feel pride in the ability to recognize their ideas in formal notation (for instance, the Commutative Property, the Distributive Property, and the statements about inverses and identities that have been explored in the seminar). Others might recall uncomfortable emotions from previous experiences in which they were presented with these laws as students.

Begin the whole-group conversation by asking if there are any of the laws that need to be clarified. Provide numerical examples and story contexts for the laws the participants are curious about. In particular, you might want to address the connections between the Commutative and Associative Properties and the work of Session 3, reordering terms and factors. See Maxine's Journal (pp. 216–220) for an example of such a discussion.

Math activity and case discussion: Factors of factors   (70 minutes)

Groups of three (40 minutes)
Whole group (30 minutes)

Because the cases in Chapter 7 present significant and complicated mathematical issues, the mathematics and case discussion are integrated. Distribute the "Math activity and Focus Questions" Sheet and let participants know they will have 40 minutes to work on these questions in small groups. Remind participants to take the time to model the questions with cubes, diagrams, and story situations. As they work, participants should also be looking for connections among different representations for the same arithmetic expressions or equations. They should also refer to the Criteria for Representation-Based Proof as they produce their arguments and examine those offered by students.

Focus Question 1 provides one story context—packaging bottles into cases— for examining factors. Encourage small groups to draw diagrams or use cubes to model the situation. You can also suggest that they create a different story context that would be appropriate for the same numbers.

Focus Question 2 asks participants to build models to show that all the factors of 48 are factors of 192 and all the factors of 81 are factors of 324. These models of specific numbers should provide the basis for the work on the general case in Focus Question 3.

Focus Question 3 requires that participants make arguments supporting the generalization, if a is a factor of b and b is a factor of c then a is a factor of c. The work on Focus Question 1 involving a specific context and Focus Question 2 using specific numbers is the foundation for this question.

Once participants have worked through their mathematical analyses of this generalization, they reflect on the student thinking in the cases for the remaining Focus Questions. Focus Question 4 invites participants to work with the cube models used by students in Cases 31 and 32 to create arguments that will serve as proofs of the generalization.

Focus Question 5 provides a similar opportunity to examine the methods students use, this time in Case 33, and to determine how to extend the student arguments to form a proof of the generalization.

Focus Question 6 provides the opportunity for participants to discuss the difference between an argument that meets the Criteria for Representation-Based Proof and a metaphor offering a mental image that might capture some elements of the situation but not be sufficient as proof.

Focus Question 7 concentrates on Case 34. Participants should identify the various generalizations present in the case and examine the use of examples and counterexamples that students raise.

Remind participants that it is unlikely they will have the time to discuss all of the Focus Questions. With 10 minutes remaining for the small-group work, let participants know that the whole-group discussion will be based on Focus Questions 2, 3, and 4.

Begin the whole-group discussion by soliciting participants' models to show that factors of 48 are factors of 196, that factors of 81 are factors of 324, and that if a is a factor of b and b is a factor of c then a is a factor of c. Include all of the different representations generated by the group. Once the different ways of representing and modeling the situation have been described, invite the group to find connections among the representations by asking questions such as "How is this component of the model using cubes present in this diagram?" or "How does this element of the story context appear in the cube models?" See "Maxine's Journal" (pp. 222–231) for an example of such a discussion.

You should also use the time to discuss the notation $a \mid b$. Some participants might note that, in fact, the notation makes the generalizations easier to express if $a \mid b$ and $b \mid c$ then $a \mid c$. You can reinforce that notion by adding comments such as "Once you understand the need for a notation—that is, once you have developed the idea—then the symbol itself is a gift, something that makes your work easier. It is only when the symbols have no meaning for you that they cause confusion."

After all of the ways of making sense of the generalization posed in Focus Question 3 are explored, refer to the cases and discuss the models students used in Cases 31 and 32. Invite participants to act out the model and then ask how to extend this argument to create a more general proof.

Introducing the Notation $a \mid b$

The symbol "$\mid$" is likely to be new to many participants. $a \mid b$ is read "a divides b." The term *divides* used in this context implies dividing equally, that is, resulting in a whole-number quotient with no remainder. For example, $4 \mid 32$ and $5 \mid 150$ are true statements. The terms *factor* and *multiple* can be used to write statements that $a \mid b$ imply. That is, if $a \mid b$, then a is a factor of b and b is a multiple of a. In the examples given, $4 \mid 32$ implies both that 4 is a factor of 32 and that 32 is a multiple of 4. Likewise, $5 \mid 150$ implies 5 is a factor of 150 and 150 is a multiple of 5.

Statements from RAO Chapter 7, such as "Because 2 is a factor of 6 and 6 is a factor of 18, then 2 must also be a factor of 18," can be rewritten as "Because $2 \mid 6$ and $6 \mid 18$, then $2 \mid 18$." The more general statement related to this example can be expressed as "If $a \mid b$ and $b \mid c$, then $a \mid c$."

Break

(15 minutes)

Case discussion: Teachers' moves

(60 minutes)

Small groups (35 minutes)
Whole group (25 minutes)

This small- and whole-group discussion provides an opportunity for the participants to focus on the actions of the teachers in the cases and to consider the impact of these actions on their students' thinking. Distribute the "Examining teachers' moves" Sheet and then ask participants to read over the assignment.

You might want to clarify the term *teachers' moves* so that participants understand that this term includes math tasks that teachers pose, questions teachers ask, classroom structures teachers set up, what teachers pay attention to, and other pedagogical decisions. This activity invites participants to identify specific actions of the teachers in the cases and to analyze those actions to consider their impact on students. One component of this work will involve participants in making inferences about the agenda the teacher has for her students. When participants offer such inferences, be sure they identify statements to support their ideas.

As you observe small groups, be sure to help them stay focused on the situations in the cases. While this time is for talking about what teachers might do or say, comments should still be grounded in the cases. Ask questions so that participants do not just make general comments, such as "She asked thoughtful questions," but are offering specific examples, such as "When she asked _____, students responded by _____."

During the whole-group discussion, solicit examples noted by participants. Examples from Alice's cases might include:

- Alice offered a variety of statements and invited students to make arguments for each. This strategy provided entry for all students and also had the potential for stronger students to look for the most general case.
- Alice noticed that different students were thinking about a similar idea, and she found a way to bring that idea to the whole class—this proved that she was listening very carefully.
- Alice used her mathematical knowledge to generate an example she thought would engage all of her students.
- Alice had a plan or an agenda for the mathematical ideas she wanted her students to work on.

When such examples are offered, be sure to have the participants cite the line numbers to keep the conversation grounded in the cases.

Homework and exit cards

(5 minutes)

Whole group

Inform participants that the reading for the next session is not a set of cases but rather an essay that will provide the opportunity for them to reflect on the ideas of the seminar.

Introduce Chapter 8, "The World of Arithmetic from Different Points of View," by Stephen Monk. Give participants a few minutes to look over the Table of Contents of the essay. Let them know the writing assignment is designed to capture their reactions to the essay and will be the basis for discussion at the next session.

Advise the participants that they might find terms in the essay that are unfamiliar. Suggest that they make a list of such terms and continue reading. The discussion at the next session will include an opportunity to clarify these unfamiliar terms.

Exit-card questions

- What mathematical ideas were highlighted for you in this session?
- What questions about teaching and learning did this session raise for you?

Before the next session...

In preparation for the next session, read what the participants have written, and write a response to each one. See the section in "Maxine's Journal" on responding to the sixth homework. Make copies of both the papers and your response for your files before returning the work.

You should also read over the statements of generalizations written by the participants to get a better sense of how they express generalizations that they observe. Looking at these throughout the seminar sessions should help you track how your participants are advancing in their abilities to express generalizations in terms of words or notation.

Math activity and Focus Questions: Factors of factors

1. Consider this context: I have a stack of cases of sodas; each case holds 24 bottles. If there is a total of 384 bottles, are all the cases full? How do you know? If there is a total of 362 bottles, are all the cases full? How do you know? How does the mathematical term *factor* relate to these questions and your responses? Draw diagrams or build cube models to illustrate these problems.

2. 48 is a factor of 192. Are all the factors of 48 also factors of 192? 81 is a factor of 324. Are all the factors of 81 also factors of 324? Draw diagrams or build cube models to illustrate these problems.

3. Is there a generalization you could make from the examples in Focus Question 2? What would it be? What kinds of arguments can you make to explain how this generalization is always true?

4. Cube models are used first by Fran and Holly in Case 31, and then by Allan and Ben in Case 32. Work through those models to understand what students are discovering. Do you think these methods constitute a proof of the general statement? Why or why not? If not, how would you modify each method so it does represent a proof?

5. In Case 33, the teacher, Jean, challenges her students to work on finding ways to show that something is always true. In response, her students offer a variety of arguments, such as using the hundreds chart, referring to dollars and quarters, skip-counting, and drawing arrays. Work through each method to understand what students are doing. Do you think these methods constitute a proof of the general statement? Why or why not? If not, how would you modify each method so it does represent a proof?

6. In Case 31, line 45, Ben compares the work with factors to moving cards when playing solitaire. Does this image offer a strategy to prove the generalization the class is discussing? Why or why not?

7. In Case 34, what are the generalizations this class is working on? How are they related? How are they related to the issue that Anthony brings up? In the case, students use numerical examples as either illustrations or counterexamples of the generalizations. What kinds of arguments can you make to show that these statements are or are not always true?

Focus Question: Examining teachers' moves

The teachers in these cases make a number of decisions and strategic moves to draw their students' attention to particular mathematical issues. Some of these decisions are seen in the questions they pose to the class, and some are seen in classroom activities they set up. Examine the cases to identify specific examples of these teaching decisions or moves.

For each example, identify the teacher's move: What has the teacher done?

Then address the following questions in detail:

- What can you infer about the teacher's agenda for her students?
- How does the teacher's move connect with where she thinks her students are?
- What do you think the teacher is trying to accomplish?
- What is the impact of her move in terms of the ideas students engage with?

Eighth Homework

Reading assignment:

Read Chapter 8, the essay, "The World of Arithmetic from Different Points of View," by Stephen Monk.

Writing assignment: Reflecting on Chapter 8

Pick three points in the essay that you found particularly interesting. Write about each of those points by explaining what made it interesting and detailing how you connect to those ideas.

Pick two points in the essay that you found confusing or with which you disagree. Explain each of these points by commenting on what was confusing or how your experience differs from what is suggested.

Bring this writing to the next session to support discussion of this chapter.

Wrapping Up

Mathematical themes:

- What does it mean to state and justify a generalization?
- What does mathematical justification look like in an elementary classroom?
- What are the connections between various representations (such as diagrams, drawings, cube structures, or story contexts) and algebraic notation?

Session Agenda

Discussion: The World of Arithmetic from Different Points of View	Small groups Whole group	35 minutes 25 minutes
Break		15 minutes
Different ways of "knowing"	Small groups/Whole group	75 minutes
Closing	Whole group	30 minutes

Background Preparation

Read

- the Casebook, Chapter 8
- "Maxine's Journal" for Session 8
- the agenda for Session 8

Work through

- the two math activities suggested for Session 8 (pp. 270–273) and choose one

Posters

- Criteria for Representation-Based Proof
- Ways of Knowing (p. 270)
- "Word Relating to Algebra Experiences" from Session 1

Materials

Prepare

- assignments and exit cards
- addressed and stamped envelopes

Duplicate

- "Final Reflection Questions" (p. 274)
- "Evaluation Form" (p. 275)

Obtain

- calculators

Maxine's Journal

December 3

We had our final session last night. At the beginning of the seminar, eight sessions seem like a long time, but it sure does go by quickly. We covered a lot of territory during that time, and last night's session was intended to wrap things up. A third of the time was dedicated to issues raised by the essay participants had read for homework. The essay was intended to state explicitly the key ideas that have been at the heart of our work, so our discussion provided an opportunity to solidify some of those ideas.

Now at the end of the seminar I also wanted to push further on ideas about proof and justification in mathematics. We explored another generalization and considered different ways of knowing that a general claim is true.

We also spent time reflecting on what happened in the seminar from participants' point of view.

Discussion: "The World of Arithmetic from Different Points of View"

For homework, I had asked that participants read the essay "The World of Arithmetic from Different Points of View." Then they were to write about three points they found interesting and two points they either found confusing or disagreed with. By completing that writing, they were prepared to move right into the meat of the essay. In small groups, they had an opportunity to talk about what they found most interesting and sort out what was confusing.

Small groups

The small-group discussions were quite energized. I had the feeling that participants were excited by some of the ideas in the essay and felt good about having had experiences that allowed them to interpret the essay. They also really appreciated the comments about the history of mathematics, especially about how long it took for 0 and negative numbers to become accepted as numbers.

I listened in on the small groups to help determine which ideas I wanted to emphasize in the whole-group discussion. I knew it would be important to spend a good amount of time on mathematical justification. In fact, I was prepared to have the group work on a new generalization in order to highlight some of the points made explicit in the essay. Before we got to that, I wanted to give participants a chance to bring up some of the other points that struck them.

In almost every group, I heard comments about the analogy of knowing your way around a neighborhood. With all the talk about whether math class should focus on memorizing procedures for calculation or should attend more to conceptual understanding, the analogy seemed like an endorsement of participants' commitments. When I opened the whole-group discussion, this was the first idea participants raised.

Whole group

James started by elaborating on the analogy. "Of course you need to know how to get from home to school and back. It might be that you know just one way that someone showed you. But if you know the neighborhood and it is a sunny day, you might take the more scenic route, and if it is a rainy day, you might take the more direct route, and if you have to stop at the library, you'll know how to take that detour and still get home."

The thing is, even though everyone really liked the analogy, I was not sure to what extent it was helping participants think more deeply about learning mathematics. I acknowledged what James was saying but asked him to relate it specifically to mathematics. He said, "Of course, students need to learn to calculate. But if they stay focused only on that one procedure, it is pretty limited. If they can look at the operations from all these different perspectives, then our students will not only know how to calculate, but they'll also know so much more. They can decide if they want to use a different procedure."

Madelyn brought in a different perspective, "I'm also thinking about having different models to use when you're trying to figure something out. Like in the cases from last time, when students were working on factors of factors, they could think about it using cubes or arrays or skip-counting."

Vera added, "It helped *us* to look at it all those different ways."

Charlotte carried Vera's comment further, "I'm thinking about the day we worked on the Distributive Property. We looked at that one problem talking about students and pencils, and we saw how the parts of the problem showed up in an array. We looked at how we did the calculation, and then we looked at how it is written algebraically. I know there's more to learn about that neighborhood, but I sure do understand the Distributive Property a lot better by looking at it through all those different lenses. It also gives me confidence that I'll be able to interpret algebraic notation in other settings."

Phuong moved the discussion to the next point. "When the author wrote about the importance of making generalizations explicit, it helped me recognize an idea that has been in my head throughout the seminar, but until now I wasn't sure what it was. Why do we go on and on about something that seems so obvious? Like why did we spend so much time talking about commutativity, when we all know it already, and so do students—or multiplying by 1?"

Phuong's energy was challenging, as though she did not really agree with the author's point of view. She seemed to be saying that it wasn't useful to spend so much time on making these ideas explicit.

Grace disagreed with Phuong. "But I don't think it was so obvious to students. That is why we were looking at it."

Phuong said, "When we looked at the DVD clip of those second graders, they all understood that you can switch the numbers when you add."

M'Leah pointed out, "You can't always tell from a DVD what's going on with students who are quiet. It might not be at all obvious to them. Also, when the teacher asked if they agreed that switching the numbers would always work, a couple of students said they weren't sure."

Denise brought in another example, Kate's case from Chapter 2. "I'm thinking about the discussion we had about a case in the beginning of the seminar. Kate was talking to her class about subtracting an amount from one addend and giving it to the other. At first, two students were talking and it seemed really obvious to them. It was only after the teacher told them to work with small numbers and show it with cubes that other students got involved."

M'Leah said, "I want to go back to that DVD segment of the second graders. They could all say that 20 + 5 and 5 + 20 give the same answer, and they could show it with cubes. But remember, the discussion didn't end there. They looked at whether it works for big numbers, and then they talked about what happens when you switch the numbers for subtraction. I don't think any of those students were bored with that."

All of these points felt important to me, but I wanted to highlight this particular one. I said, "M'Leah's example illustrates another important point. One reason that it helps to make ideas or strategies explicit is that doing so makes the limits clearer. By having the discussion about whether you can switch numbers around for addition, you're then in a position to talk about whether you can switch numbers around for subtraction. We saw in a few of the cases that some students will over-generalize; some will conclude that you can switch numbers around, period. If you're explicit about these ideas early on, then you can refer back to them later when things get more complicated."

Yanique picked up on what I was saying. "I'm interested in what the author said about students having a more clearly articulated sense of the operations and how that will serve them well later in school."

Charlotte said, "I'll tell you, it really helped *me* to have those discussions about what the operations do and then look at the algebraic notation. I could start to see the algebra as statements, or descriptions of situations, a way to say what we were talking about. Like when we looked at the algebraic statement of the Distributive Property, each side of the equal sign was a different way of describing the same situation."

Madelyn pointed out that you don't have to wait until students are learning algebra to see that it is useful. "I have been paying attention to the ideas my students use when they are working on double-digit computation. We stopped and had a discussion about whether order matters when adding and whether order matters when subtracting. Once we had that discussion using smaller numbers, we could refer back to it when they shared their computation strategies."

I said, "The author of the essay talked about having 'portable tools.' What do you think he meant by that?"

Madelyn said, "That is what I mean. Once we talked about whether order matters, and we talked about it with small numbers, we could use that idea in another context."

A light bulb now seemed to go off for Denise. "Ohhhh. So when we're surprised that students seem to know something in one context but not in another, it might be that they never explicitly thought about the idea. Maybe what we need to do is go back to the context in which they have it right, make that idea explicit, and then show them how the same idea comes up in the other context."

It seemed that Denise had in mind a constellation of examples, but I chose not to pursue it further. She was making an important point, but there was more we needed to get to.

Jorge brought us to the main issue that I wanted to work on, though I was a little surprised by the way he brought it up. He said, "I think the author too quickly dismissed a statistical approach. The last time I checked, statistics is a branch of mathematics."

I asked, "Can you say more? I'm not sure what you mean."

Jorge continued, "It seems to me that the statistical approach can be as convincing as a logical approach. I take exception to the notion that a statistical approach is entirely invalid. It seems to me that if a few significant examples work, then we are moving toward a point where we can accept a generalization."

Wow. I was surprised by the energy behind Jorge's words. I think he realized that he figures out a lot of mathematics empirically and feels that part of the process was being disparaged. I had already decided it was important to spend some time on this issue of justification. I wanted the group to explore the role of empirical evidence in learning mathematics and the difference between that and proof. However, I also wanted to help Jorge, in particular, come to another view.

In the moment, I was discouraged that Jorge was raising this issue at the end of the seminar. I had thought that, throughout the seminar, we had been working hard to think about what it means to prove a general claim. Now that I am writing, I realize that, although the contrast between testing examples and mathematical argument arose periodically, it was never stated as starkly as in the essay.

I addressed the whole group, "Let's set aside what's stated in the essay for a moment, and let's just think about the difference between empirical evidence and the proofs we have been using based on models. What do you see as the difference?"

Lorraine said, "It's like what we said about the neighborhood. All those models gave me a different lens to look at the idea. I feel like I understand the idea a lot better by using different models."

Charlotte said, "I just opened up the Casebook to the first page. There's a justification about adding even numbers. If you have a bunch of pairs and a bunch of pairs and you put them together, you have a bigger bunch of pairs. It is so clear that has to happen no matter what two even numbers you start with. You'll always end up with an even number; it is always going to be that way, without a doubt. That seems really different from testing a few, or even a lot, of examples. The examples don't show you why it has to be true."

Yanique added, "I remember Maxine's example from the first session, too—about adding prime numbers. If you add two prime numbers, will you always get an even number? That really struck me. You can keep getting more and more evidence, adding more and more pairs of primes, and always get an even number. If you don't happen to try a 2 with a different prime number, you'll always get an even number. But you'd come to the wrong conclusion if you think that is enough evidence."

Jorge said, "But a lot of times it is right. You don't want to throw the baby out with the bath water."

Although people were trying to convince him of another point of view, Jorge was sticking to his guns. I wanted to make sure he did not feel backed into a corner in this discussion, and while I wanted him to see the difference between a statistical argument and a mathematical argument based on reasoning, I also wanted to make sure he knew I was listening to him.

I said, "Jorge, you're right that we don't want to throw the baby out with the bath water. So, let's think about how lots of examples *can* help us to draw conclusions."

June pointed out, "The essay talks about that, too. Students learn to call something a dog by making generalizations. And they learn to make a word plural by adding an *s*."

I asked, "And how do we use examples to learn mathematics?"

Denise said, "By looking at examples, it is like we can formulate conjectures. Then those are the things we need to prove."

Denise was making an important point that I wanted the group to pay attention to, so I asked, "What do you think about Denise's point?"

Antonia said, "When we were working on factors of factors, even though we had read the cases before class, I didn't really understand all the math. So,

when we did the math for ourselves, we first checked it with particular num- | 195
bers." Antonia flipped through her notebook to find her work from Session 7.
"We saw all the factors of 48 were also factors of 192, and all the factors of 81
were also factors of 324. Anyway, at first I was surprised that it was all coming
out that way. Then I was pretty sure that it would always come out that way.
So I guess you could say I was pretty sure of my conjecture." | 200

Antonia was giving us a helpful illustration, so I explicitly restated
Denise's point and then brought us back to Antonia: "The examples help you
formulate a conjecture—something you think is probably true but hasn't yet
been proved. So, Antonia, what happened after you were pretty sure of your
conjecture?" | 205

Antonia said, "Then we looked at all those different ways to prove it, and
it was pretty neat."

Before I could continue with Antonia's example, Jorge interjected, "Let's
say there's something your class is working on and they have lots of examples
of it. Isn't it better for your students to know that it is true instead of feeling | 210
unsure because they can't test all examples?"

I said, "Actually, from a mathematical point of view, the skepticism is war-
ranted until you actually prove it. That is what characterizes mathematics—
that there are things you can prove for an infinite number of cases without
testing all cases. The essay points to another example of this. We recently read | 215
Jean's case from Chapter 7. Let's look back at it. When students looked at lists
of factors of 100, 200, 300, and 400, they made lots of observations."

Kaneesha pointed out, "Some of them are right and some aren't."

I said, "Yes, some of the conjectures turned out to be true and some didn't.
But, they made all their observations from what seemed like quite a bit of | 220
evidence. So, how do you decide whether one of these observations applies to
all multiples of 100? It is true you can find all the factors of 500, 600, 700, and
keep going, seeing if the observations still hold. But mathematicians also have
another way to know."

Jorge protested, "But you'll never get anywhere if you have to prove abso- | 225
lutely everything when you're teaching mathematics."

Throughout this discussion up until now, as I was offering examples or
working to bring out the points from the group, I was thinking about the
importance of understanding the difference between mathematical proof and
empirical evidence. At this point, I started to realize that Jorge was now think- | 230
ing from the practical standpoint of a teacher. It is important to call upon various modes of "knowing" when you are learning or teaching mathematics.

I said, "Yes, Jorge, there are other considerations when you're learning
or teaching mathematics. Sometimes it is appropriate for students to accept
mathematical truths on the authority of the teacher or the textbook. Sometimes | 235
it is appropriate to go with your hunch, based on examples, even if you can't

prove it. But it is important to know the difference between those ways of knowing and having a mathematical argument based on logic."

Once I acknowledged this, I saw that Jorge's affect had changed. Now that he did not feel as if empirical evidence was being disparaged, he was ready to think about these different ways of "knowing."

Jorge had surprised me by the way he brought up the issue, but I realize that it was probably helpful to many members of the seminar. I had intended to have a discussion about mathematical justification anyway, but the starkness of Jorge's position brought some points into sharper focus.

Different ways of knowing

Now, I initiated the discussion I had planned ahead of time. "Let's try another conjecture, just to explore these different ways of knowing." I wrote the following list on the board:

- Accepting on authority
- Trying it out with examples
- Applying mathematical reasoning based on a visual representation or story context
- Proving using algebraic notation and the Laws of Arithmetic

I continued, "I'm going to state a generalization, and then we are going to think about these different ways of knowing that my claim holds for all numbers. I want you to actually spend some time with each of these ways of knowing in order to see the distinctions. So first, I'll state a generalization, and I want you to think about what it is like to accept it as true, based on my authority. Here's the generalization: When you multiply two numbers, you can cut one number in half and double the other, and the product will stay the same. I want you to write the statement down and then write about whether you believe it and why—and what it feels like to accept it that way."

I wrote the statement on the board and then waited about 2 minutes before I continued. "OK. We'll talk later about what you wrote down. Now, I want you to test it out with some numbers. You can work with the people who are sitting near you. If you have a calculator, you might find it handy."

The groups I saw started out with small numbers: $4 \times 6 = 2 \times 12$, $6 \times 5 = 3 \times 10$. Then they used larger numbers that are still convenient: $12 \times 15 = 6 \times 30$, $14 \times 40 = 28 \times 20$.

After a few minutes, I stopped the group. "What are you finding?" I asked.

June said, "It seems to work every time."

M'Leah said, "Wait a minute. I think I found one that does not work. What about 11×15?"

I asked, "What does the generalization say about 11×15?"

James said, "We need to check 5.5×30. Does that give the same answer as 11×15?"

M'Leah said, "Oh, yeah. We can still take half of 11, even though it does not give you a whole number." She did some scribbling on her paper. "They both come out to 165. So it does work."

Charlotte said, "I know that some of my students occasionally use this as a strategy for multiplying. Sometimes it works, and sometimes it does not."

I said, "We need to be careful about what it means to 'work.' Are we thinking that it works if it makes the multiplication easier? Or are we thinking that it works if the generalization is true that the two products are the same?"

Grace said, "Sometimes this is a convenient strategy to use, and sometimes it is not. But the generalization can be true, even if it involves messy numbers."

I was ready to go on. "At this point, the evidence is supportive of the idea that the generalization is true. But maybe we have not hit upon the examples that disprove it. Now, let's see if you can come up with an argument to show that it will *always* work. I want you to work on the last two ways of knowing on our list." Small groups began to work on developing mathematical arguments based on visual representations or story contexts or creating proofs using algebraic notation.

Once we came back together as a full group, we worked through the different arguments. Claudette started us off. "Let's say I have 12 bags of apples, and each bag has 15 apples in it. The total number of apples is 12×15. But now I have to carry them all to the car, and so I take one bag in each hand, and I need to make 6 trips. Because I take two bags each trip, the number of trips is half the number of bags, but the number of apples I take each trip is double the number in a bag. That means the total number of apples is 6×30."

I said, "OK. Claudette has given us a good context to use. To help us get the context in our heads, she used the example of 12×15. Now, let's see if we can use that context to think about the generalization without using the specific numbers."

Denise suggested, "We can use Claudette's story, but we can talk about a and b instead of 12 and 15. Say the number of bags is a, and the number of apples in a bag is b. The total number of apples is $a \times b$. Then, when I carry the bags to the car, I take 2 bags at a time. So the number of trips I take is $\frac{a}{2}$ and the number of apples I carry each trip is $b \times 2$. So, the total number of apples is $(\frac{a}{2}) \times (b \times 2)$."

I wrote some of Denise's notation on the board and asked, "Does this argument make the case for all numbers?"

Madelyn said, "It goes beyond testing examples. Theoretically, you can have an infinite number of cases covered by Claudette's story, except that there aren't an infinite number of apples in the world. Still, it provides an image for what's going on when you cut one number in half and double the other."

M'Leah protested, "But it does not include my example. What about 11 ×
15? It doesn't make sense to talk about 5.5 trips to the car."

Claudette said, "Right. My story works only if you have whole numbers.
So, at least one of the factors has to be even."

I suggested we look at some other arguments and then get back to this idea
about whether it works only if at least one factor is even.

Grace said, "I looked at arrays." She held up an array of cubes, and then
showed how she broke the array into two parts and rearranged the parts.

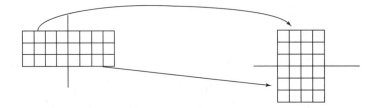

Grace explained, "When I broke the array into two parts and put the parts
back together, I ended up with one dimension cut in half and the other dimension doubled. But the total number of cubes stayed the same."

Before I could ask a question about it, Claudette said we were still in the
same situation. "You can do that with any array, as long as one of the dimensions is an even number. But we still don't have M'Leah's case covered. Maybe
it works only when one of the numbers is even."

M'Leah said, "But my example worked. 11 × 15 = 5.5 × 30. I think we need
to find a representation that isn't just whole numbers."

I said, "There's something I want to point out here. So far, our arguments
only work for whole numbers, and even then only when one of the original
numbers is even. So we need to question how general our generalization is.
We're not sure if it is true for fractions. But we have an example that leads us
to believe that it might be."

Phuong said that she has an argument. "It is like Grace's argument with
arrays, but it does not use cubes and does not require whole numbers." She
came up and drew a rectangle.

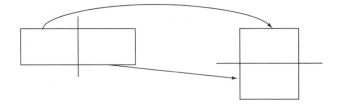

Phuong explained, "Just like Grace did, I cut one rectangle in half and rearranged the parts. The area of each rectangle is the same because they are both
double the same thing. But to make the second rectangle, one dimension was
cut in half and the other dimension was doubled."

Wrapping Up

M'Leah said, "That's it! That covers fractions, too. The length of a side of the rectangle can be any number."

Actually, the rectangle demonstrates the truth of the generalization for any positive number, but I decided not to push that right then. Instead, I chose to take a few minutes to talk about the three ways of knowing we had worked on so far. I said, "At first, I asked you to take a moment to write about what it felt like to accept the generalization as true on my authority. I would like to hear what you wrote."

Kaneesha said, "When you just said it was true, I felt like there wasn't any point to explore it. It took all the excitement out of the problem. I was just supposed to accept it."

Yanique said, "I didn't believe her. Maxine said it was true, but that didn't mean it was. I didn't wait to start testing examples or drawing diagrams, so I started right in."

Charlotte said, "I could not make any sense of the general statement when it was just there in words. I needed to try some examples to see what the generalization actually said."

I said, "OK. So Charlotte has suggested another way to use examples. She could not interpret what the generalization said until she chose some numbers to see what was going on."

Yanique then continued to talk about her process. "Once I tested it out with a few examples, I couldn't wait for you then, either. It seemed like it was true, so I needed to draw some diagrams to see what was going on. I drew a diagram like Phuong's. But I really like Claudette's story context, too. Even though it is about whole numbers, it helps me think about multiplication in another way. The rectangle helps me think about multiplication as area, and Claudette's story is about groups of things. You don't need the apples. If you have a bunch of equal groups, and if you pair up the groups, you end up with half the groups with double the stuff in each. That makes me think that it works for M'Leah's numbers, too. You can still think about half a group of double the stuff."

Grace said, "I wanted to go along with what Maxine said. I accepted the generalization on her authority, and then I tested out examples. It wasn't until I started working with the array that it became obvious that the product will always be the same. I see that you need to think about a rectangle more generally, not just with whole numbers."

Mishal, who had immediately gone to an algebraic representation when I first asked everyone to come up with a mathematical argument, had been very patient. Now I asked her to show us her way of thinking about it.

Mishal explained, "Actually, it is like what Denise said about Claudette's story. I just did this." She wrote:

$$\tfrac{1}{2} \times a \times 2 \times b = a \times b \times \tfrac{1}{2} \times 2 = a \times b \times 1 = a \times b$$

Mishal turned around and thought she was done, but clearly many members of the group needed more explanation. June said, "I don't get it."

Denise suggested that if Mishal added parentheses, it would be easier to see. Once Mishal did that, her equation looked like this: 390

$$(\tfrac{1}{2} \times a) \times (2 \times b) = (a \times b) \times (\tfrac{1}{2} \times 2) = (a \times b) \times 1 = a \times b$$

Mishal explained, "Say you start with $a \times b$. You need to show that if you cut a in half and double b, you end up with $a \times b$ again. So first I multiply a by $\tfrac{1}{2}$ and multiply b by 2. Then I change the order of the factors to put $\tfrac{1}{2}$ and 2 together, because they equal 1. And then I end up with $a \times b$." 395

I told Mishal that this was a good proof, but I pointed out that if someone really wanted to be picky, there are a few additional steps to take. When she went from the first expression to the second, she was both changing the order of factors and changing the way they are grouped, applying both the Commutative and Associative Laws at once. 400

Denise then said she had another way to prove the generalization. "I'll use different letters. But I'll start by saying the first number is $c \times 2$. I know that any number can be written that way. It just means that half the first number is c. So, my proof looks like this." Denise came to write on the board.

$$(c \times 2) \times d = c \times (2 \times d)$$ 405

Denise explained, "We know that is true by the Associative Law of Multiplication. You multiply two numbers. Then, because c is half the first number and $(2 \times d)$ is double the second number, it proves that you get the same product."

Lorraine said, "I like Mishal's proof better."

Antonia said, "So do I." 410

One of the things that I have been noticing as I work with elementary teachers on algebraic notation is that, sometimes, the notation is too terse to be comprehensible. People who are already fluent with the notation are likely to prefer Denise's proof because it seems more direct. However, the notation is so packed, it is difficult for novices to parse. Participants like Lorraine and Antonia 415 will often find it easier to see their ideas in notation that is less concise.

Jorge said, "I just noticed something else. If you look at Phuong's rectangle, you can see that the generalization will work for any positive number. When you look at Mishal's proof, you can see that it works for negative numbers, too."

Closing

Reading and discussing the essay was one way to think back on the work we 420 had completed over the last eight sessions. I still want everyone to reflect on his or her experiences in a more personal way. For homework (yes, there is still one more assignment after the last session), I asked that everyone review their portfolios and respond to questions about what they learned in the seminar.

For now, it was time for us to come together to say good-bye. I reminded the group of the words or phrases they offered at our first meeting to characterize their experiences with algebra. Now I wanted them to share a little about what has changed for them. Following are a few thoughts that were expressed:

LOURDES: This seminar has given me confidence to allow students to converse and build representations. It has helped me to know what kinds of questions to ask to support these conversations.

RISA: I am feeling more comfortable about taking the time to really pursue questions. I am OK with coming back to a discussion on another day and to let students keep working with ideas.

CHARLOTTE: Algebra is not scary to me any more.

ANTONIA: I now have a better idea of what I am evaluating on our report cards when it asks for algebraic sense. I now see algebra as relationships between numbers and operations.

DENISE: I have a better understanding of how people learn. Even though I did well in math, I wish that I could have learned it this way. These are conversations that I want to have with my students. But I am frustrated with the demands of our curriculum and the time we have to teach it.

M'LEAH: I passed algebra, and my teacher thought I was a talented student. But I realize now that I did not really understand algebra, and I know my teacher did not realize that I did not understand it. Now I have a better idea of what to listen for to know what my students know.

JUNE: I really have come to understand and value the question, "Why?"

KANEESHA: For the first time, negative numbers made sense to me. This was because of the work we did with multiple representations. I am now thinking about "full" representation with my students: pictures, cubes, stories, numbers, symbols, and diagrams.

JAMES: I learned what it means to make and prove generalizations, but I am frustrated about whether or not I'll have the time to do this.

LEEANN: I hated math, and I did not like teaching it. For the first time, I feel like I am not stupid when it comes to math. I realize that I can figure it out, and now I am wishing that I could have pursued a career that involved more math. I didn't because I didn't think I could do it. Now I am wondering if I shortchanged myself.

JORGE: I am seeing that I can be a teacher and a learner at the same time in the same day.

YANIQUE: I see the power of doing problems together. We all helped each other to see things in new ways and to see the bigger picture. It has been important for my understanding and I want to support this kind of collaboration for my students.

I thanked the group for sharing these thoughts. I also told them how much I appreciated their openness, their confusions, and their insights throughout the seminar. I have learned a great deal from working with them. I reminded them that their final reflections will help me better understand what has happened throughout these past months.

Final reflections

December 17

By now, I have received the final reflections from my RAO seminar participants, and it is time for me to write my own final reflections. As I read through what they wrote, I get a fuller picture of the comments they made in the last session.

Leeann summarized what worked for her in this course:

On an emotional level, I was nervous and felt like I might not have enough algebra background to understand this seminar. I am happy to say that building models made most of the algebra accessible to me.

Several participants wrote about understanding algebraic notation differently or being able to use it with greater fluency. Some said that their understanding of how variables can be used had been confined either to writing function rules or to finding unknowns. Using variables to express generalizations was a new, or as Lorraine said, a "forgotten" idea.

Lorraine wrote:

I learned how to write an algebraic statement. I had forgotten that you could make statements like $a + b < c + d$. I had also forgotten that you could set up a situation with language like

If $a + a = b$, then $a + (a - 1) = b - 1$.

June wrote:

When I thought of algebra, I recalled it having letters that stood for a specific number. a + 5 = 7; a = 2. These letters were part of an equation and I knew it would balance.

Now I know that the letters can stand for a set of numbers that have a relationship to each other: a + b = c can mean 2 + 5 = 7 or 4 + 10 = 14 or 10 + 25 = 35. I have found that equations can be modeled by drawings, geometric representations, numerical sentences, and when given in a context or my giving a context to "it," I have the greatest understanding.

Through all of her struggles with algebraic notation, Charlotte came through the other end. She wrote:

One idea that has changed is my viewpoint of the use of symbolic notation. I did not really see its "usefulness" before this, but now I can see how it allows us to capture the explicit points of a generalization in a compact and elegant way. I can almost (not quite) make an analogy to the power of a perfect word in a poem.

Grace does not see the poetry the way Charlotte does, but she feels more fluent and can appreciate the notation's usefulness:

I can generate algebraic notation pretty easily now, although it does not satisfy me the way a visual representation does with regard to justification. However, I can see why it does satisfy some people. I see the connection to logic. I see it is useful—a useful discipline.

Quite a few people wrote about how their understanding of the Laws of Arithmetic has changed.

PHUONG:

I never dreamed that the Laws of Arithmetic could be shown in models. This was a wonderful "aha!"

MADELYN:

I'm much clearer about the importance of logic and also the role of the Laws of Arithmetic within justification. I used to find them sort of foolish and irritating. Now I see that they have a useful and proper place, and I can keep them there and not be irritated by them. I learned that the Laws of Arithmetic are about the operations, not the numbers—about the operations in a particular realm of number. That distinction has, I think, made me less irritated by them.

JAMES:

I know now that the Laws of Arithmetic have a greater importance than I have ever given them. Knowing the great foundation they give to higher level mathematical thinking makes them more meaningful. I feel I understand them, but I want to internalize them with context.

VERA:

Before the seminar, I was thinking of algebra more in terms of the patterns and functions part. I probably blocked out all the parts related to the Laws of Arithmetic because I never really internalized anything about these laws in a meaningful way. They were always just "out there" in a theoretical way. Those darned laws, which have always seemed to be somewhere between bothersome and cumbersome, carry more meaning and significance for me after our work in the seminar.

Most participants wrote about the generalizations we worked on in the seminar.

JORGE:

I have no recollection of work in generalizations and justifications in high school or college algebra.

ANTONIA:

I'm seeing the big picture about algebra more—that it is about generalizations, justification, and representation; that justification is derived from logical arguments. I have a much clearer idea of the role of representation—as a vehicle to communicate ideas about justification. These words mean something to me now.

CLAUDETTE:

I have never really thought about how to make a model that shows a generalization, a model that uses variables or has the quality of infinity built into it. This seems very powerful to me.

Lourdes said that the emphasis on mathematical reasoning gave her a different sense of the usefulness of algebra.

I have learned that algebra has a function in life, not just high school. In high school it was a class. Now, I see it as a way to make sense of a mathematical question. Algebra can help make computation less of a chore because I understand why I'm doing the task. Algebra has given a more pronounced meaning to each operation and how they are different from one another. Algebra is a thinking model to me and it becomes a portable tool and it wasn't even in my toolbox before this seminar.

The work on generalizations brought participants to new ideas of what can happen in the elementary and middle-school classroom.

M'LEAH:

Prior to this class, I thought of algebraic thinking as using patterns to make predictions and using variables to express rules. This seminar has helped me see that algebraic thinking also requires students to think about why the mathematics works and test out whether or not it will always work or for which kind of numbers. Algebraic thinking includes justifying work with models. I'm learning that algebra is more than x's and y's and plugging num-

545

550

555

560

565

570

575

580

bers into equations. It includes students defending understanding with context, numbers, and models.

Risa:

> *I have a much stronger sense of what justifications look and sound like in a fifth-grade classroom. I am able to connect my own learning experiences in the seminar (moving in and out of concrete work to abstract thinking, moving in and out of confusion to clarity) to how students might interact with these ideas. The complex learning that happens during these types of discussions has huge implications on students' dispositions toward mathematics as well as their computational fluency and number sense. Another idea that was developed was the role of models in establishing justifications. I can see how students can use the models (stories, diagrams, etc.) to put their ideas into generalizable images.*

Mishal:

> *My ideas about algebra have changed in the sense that I wasn't sure if math generalizations were appropriate at all levels. I now know that students will make generalizations whether I make it a priority or not. I have to create lessons and steer students to algebraically generalize. I think it will help them make more sense of the math they are doing now and in the future.*

Denise:

> *Right now I'm thinking about the importance of connections—between different models and between models, algebraic notation, context, words, and so on. Also, there were concepts (even/odd, additive inverse, and especially the identity elements) that I used to think were "duhs" or givens. I realize that if we don't explore these concepts early on, they explode in middle and high school when we get into more formal generalizations.*

Yanique:

> *Thinking about binocular vision and the discussions about students seeing the parallels and differences between the operations as well as the number relationships were huge guiding concepts for all of the work we did.*

Although Josephine struggled with much of the mathematics in this course, she, too, moved forward and took from her work important pedagogical lessons:

> *Models are so powerful! The need to build, draw, and create them and then having the opportunity to explain them is critical to helping students connect these powerful ideas.*

These excerpts do not capture all the important points that were made in participants' final reflections. For example, many wrote about how they now understand, or are still thinking about, operations with negative numbers. However, as I started finding excerpts to record in *my* final reflections, I realized I was capturing an idea from each participant. Reviewing these excerpts gives me an overview of my class.

Participants entered the seminar with different backgrounds, dispositions, and skills, and they leave in different places, as well. I have confidence that most have moved into a new place and that their new algebraic skills will serve them well. I hope that they will bring a new dimension to their teaching—of listening for the generalizations their students make about operations and challenging their students with such questions as, Will this work for all numbers? How do you know? I have confidence that these teachers will encourage their students to consider ideas through multiple representations and examine how those representations are related.

A major goal of the seminar was for participants to understand that, once a generalization has been presented as a conjecture, there is a difference between demonstrating its truth through mathematical reasoning and checking lots of examples. I am grateful that Jorge highlighted that issue in our last meeting, even though he raised it with angry energy. I believe he felt that something that he valued was being disparaged. I think our discussion allowed Jorge and the rest of the class to see that checking examples is useful but also how it differs from developing a mathematical argument. I cannot find evidence in the final reflections of whether that idea hit home or not. If I have a chance to work with any of these participants again, or to visit any of them in their classroom, I will look for that idea.

As with any class, the learning is not complete. Having opened a door to these ideas about early algebraic thinking—listening for students' generalizations, making arguments based on visual representations—teachers are in the position to learn more than what we touched on in this seminar, and through their work with their students, to make new connections.

One thing I forgot to mention in my entry after the last session: At the end, I asked what it would be like if participants now entered a more conventional algebra class. Charlotte said she would approach the class with the expectation that she would understand and wouldn't let her instructor get away with moving on when she did not. Yanique said it would be a completely different experience for her, to approach the material with greater depth. Madelyn told us that with a new sense of where the Laws of Arithmetic come from, the activity would mean something very different to her. I will be interested to hear if anyone does enroll in such a class.

Detailed Agenda

Discussion: The World of Arithmetic from Different Points of View (60 minutes)

Small groups (35 minutes)
Whole group (25 minutes)

Organize participants in groups of 3 or 4. For this discussion, mixed grade-level groups are best. In this small-group discussion, participants will have the opportunity to share their reactions to the chapter essay and to talk over the terms they found unfamiliar. Ask that they first read one another's written comments about the essay so they can integrate everyone's thoughts into the discussion.

Suggest that small groups begin by talking about the points they found interesting, and then turn to those ideas that were confusing or with which they disagreed. Participants should also use this time to share information concerning any unfamiliar terms. Be sure they understand that they can call on you to explain terms that the members of the group are unable to clarify. Questions might arise that are unresolved in the small-group work. Assure the groups that they can bring these questions to the whole-group discussion.

Begin the whole-group discussion by asking whether there were commonalities regarding points of interest. As points are discussed, ask questions or make comments to connect participants' remarks with the mathematical themes. If participants comment on the neighborhood analogy, ask them to be specific about what it means to them in terms of learning mathematics.

Then turn to any areas of confusion or points of disagreement about the essay. Clarify any terms that are still unfamiliar and respond to the issues of confusion. If questions were raised in the small groups that were not resolved, invite participants to pose those questions for the whole group to discuss.

Break (15 minutes)

Different ways of knowing (75 minutes)

Small groups/Whole group

An important theme addressed throughout the seminar is the difference between using examples to formulate or support a conjecture and generating an argument based on mathematical logic and structure to prove the conjecture. While both are important aspects of doing mathematics, it is likely that many participants entered the seminar being comfortable examining examples to test a conjecture and having little experience with proof. Throughout the seminar, participants have worked on developing arguments using visual representations, and this subject arises again in the Chapter 8 essay.

This final mathematics activity will provide another opportunity to emphasize the difference between checking particular instances of a generalization and proving that it holds for an infinite class. Depending on the past experiences of your seminar participants, choose one of the following generalizations as the basis for this work:

1. When you multiply two numbers, you can cut one number in half and double the other, and the product will stay the same.

2. When you multiply two square numbers, the product is also a square number.

If your participants have not worked on the idea in the first generalization, you should begin with this generalization as the math activity. If most of your participants have worked on this idea in the context of their work in Session 6, then choose the second generalization as the basis for this session. Examine the descriptions of these two choices for the math activities near the end of this agenda to help form your decision. In some seminars, there may be time to work on both generalizations.

Refer to the conversation from the essay discussion to bring the points about what constitutes mathematical proof to the fore. Display the poster describing ways of coming to believe a mathematical statement is true:

1. Accepting it on authority

2. Trying it out with examples

3. Applying mathematical reasoning based on a visual representation or story context

4. Proving it using algebraic notation and the Laws of Arithmetic

Participants will work on the chosen mathematics question in all four ways and compare and contrast what each contributes to the experience of knowing. In doing this, they will examine the connections between representations, such as a diagram, and the more formal algebraic symbols for the same mathematical question.

One hour and fifteen minutes have been set aside for this activity and discussion. Because the small-group and whole-group work is interspersed, be conscious of time so that you will have at least 25 minutes for the final whole-group discussion.

To address the first way of knowing, accepting on authority, begin by writing the generalization on the board or easel. Then inform the whole group that you are telling them this is a true statement; that is, they can accept this on your authority. Ask participants to write for a minute or two about how that feels.

Next, work on the second way of knowing by inviting participants to work a few examples using numbers. Suggest that they talk to someone sitting next to them to share their discoveries.

As a whole group, discuss the examples they tried as well as their beliefs about this statement. The object of this part of the discussion is to highlight what it means to gather evidence for a claim through some representative examples. They may also find that, sometimes, what is noticed by exploring a set of examples can provide insight into the general claim.

The third way of knowing requires the use of representations to prove a claim, and the fourth way of knowing uses algebraic notation. To address each of these, pose the following task: Make an argument that illustrates this statement is true using a diagram or story context. Then examine the statement with algebraic symbols.

Announce how long participants will have to work in small groups to produce their arguments.

Call the whole group together so they can share their arguments. In the whole-group discussion be sure the variations for the arguments based on diagrams are presented. Encourage participants to examine how the representations are similar and how they are different. Ask questions to reveal how the various representations illustrate multiplication. Also, have participants determine how the argument is general and not bound to the specific numbers being used as illustrations.

Near the end of the discussion, have a participant share his or her argument based on algebraic symbols. If no small groups used algebraic symbols successfully, help participants work on this as a whole group. Ask questions about the connections between the diagrams and the symbolic argument.

Conclude the discussion by revisiting the four ways of coming to believe that a statement is true (by authority, by examples, by reasoning with a diagram based on structure, and by reasoning with algebraic symbols). Participants should use this experience to talk about the power of the last two methods. For more information on this mathematics activity, see "Maxine's Journal (pp. 256–260)."

Closing

(30 minutes)

Whole group

Use the final 30 minutes to provide closure to the seminar experience. First, distribute and explain the "Final Reflection" assignment that participants are to do as homework. This assignment has two purposes: One is to provide participants with the opportunity to examine how their thinking has evolved over the course of the seminar. Another is to provide you, the facilitator, with information about what your participants have learned. Let participants know that indicating what in their work helps them to see the changes they are writing about will be helpful.

Announce the date by which you wish the assignment to be completed and the process for returning it to you. If you are using a seminar evaluation form, distribute the form and clarify the expectations for returning the form. Provide stamped, self-addressed envelopes for the final reflection and the seminar evaluation form.

Even though participants will have time individually to reflect on the seminar experience in detail through this assignment, it is important that the seminar close with some time for the participants to express their thoughts. Remind participants of the opening day of the seminar when they offered a word or phrase to describe their experience with algebra. Then ask them to share a word or phrase that represents their current sense of algebra. You may wish to display the poster from Session 1 as a way to begin the conversation.

Poster to prepare for Session 8

Ways of knowing

1. Accepting it on authority

2. Trying it out with examples

3. Applying mathematical reasoning based on a visual representation or story context

4. Proving it using algebraic notation and the Laws of Arithmetic

---

Math Activity, Choice 1

Ways of knowing this generalization is true: *When you multiply two numbers, you can cut one number in half and double the other, and the product will stay the same.*

Accepting a generalization on authority: Simply announce this is a true statement.

Trying it out with examples: Invite participants to try out some numerical examples and gather the whole group together to see what they noticed.

Some might notice that some numbers are more awkward to use than others. For instance, 12×15 might be transformed to 6×30, producing 180 as a product. On the other hand, 11×17 would result in 5.5×34 producing a product that is correct but not easy to calculate. Reflecting on this idea provides the opportunity to talk about the difference between a mathematical principle and a computational strategy. The strategy may be more or less useful for particular numbers. However, the mathematical principle that underlies the strategy is always true and is not dependent on specific numbers.

Applying mathematical reasoning based on a visual representation or story context: One possible representation might take the form of an area model. First, consider a rectangle with dimensions m and n. The area of the rectangle is the product of m and n, $m \times n$. If I slice the rectangle in half horizontally and move the bottom half over to the right of the upper one, I make a new rectangle with the same area ($m \times n$). The dimensions of the new rectangle are $\frac{m}{2}$ and $2 \times n$. Therefore, the product of $\frac{m}{2}$ and ($2 \times n$) is also $m \times n$.

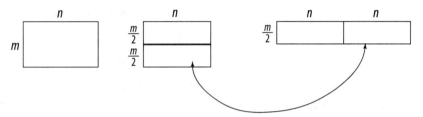

Some participants might create a similar diagram using specific numbers, illustrating, for instance, how a rectangle that is 12×15 can be transformed into a rectangle with the same area and sides 6×30. Ask questions to help the group see how the diagram solution can be made to apply more generally. As participants work with such diagrams, help them articulate the connections between the diagram and the statement by asking, "How does your diagram represent multiplication?" Also, pose questions to encourage participants to explore the generality of their argument. One such question might be, "This diagram is built with the numbers 16 and 15. How does it show that the product of any two numbers will work this way?"

Help participants articulate how the arguments satisfy the Criteria for Representation-Based Proof by questioning how multiplication is represented, how the representation applies to the general case, and how the argument supports the claim.

Proving using algebraic notation and the Laws of Arithmetic: Some small groups may have worked on a symbolic method similar to the expressions used in the diagram. One possibility might be $(\frac{m}{2}) \times (2 \times n) = (m \times \frac{1}{2}) \times (2 \times n) = m \times (\frac{1}{2} \times 2) \times n = n \times 1 \times m = n \times m$.

Another possibility is to state that the dimensions of the original rectangle are $(2 \times a)$ and b. So, $(2 \times a) \times b = (a \times 2) \times b = a \times (2 \times b)$. You might want to refer to the "Laws of Arithmetic" page to verify how the Associative Law and the laws regarding inverses and identities come into play in these arguments. See "Maxine's Journal" for Session 8 for an example of this discussion.

A possible extension of this work is to consider how this argument could be modified to apply more generally. Instead of halving and doubling, consider multiplying and dividing by 3 or 4 or k (with $k \neq 0$). Recognizing how arguments can be rethought to account for such changes in the numbers involved is a way to work on constructing the most general argument.

Math Activity, Choice 2

Ways of knowing this generalization is true: *When you multiply two square numbers, the product is also a square number.* Clarify the definition of *square number* as a number that is the product of the same two whole-number factors. For example, 9 is a square number because $3 \times 3 = 9$.

Accepting a generalization on authority: Simply announce this is a true statement.

Trying it out with examples: Invite participants to try out some numerical examples. Calculators may be useful for this work.

Bring the whole group together to discuss what participants have found and what their beliefs are about this statement. The object of this part of the discussion is to highlight what it means to gather evidence for a claim through specific examples.

Applying mathematical reasoning based on a visual representation or story context: One representation that might be used is based on the fact that every square number can be arranged as an array with dimensions that are equal. For instance, 9 is a square number because 9 objects can be arranged as a 3-by-3 square. The length and width of the array are both 3 units.

As an example, participants might begin by representing the product of 9 and 16 as 9 repeated 16 times, arranging the 16 repetitions in a 4-by-4 square.

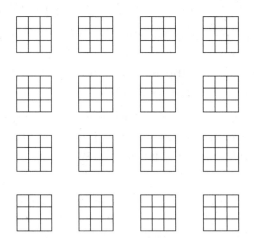

The result is a 12 by 12 array. Encourage participants using this model to talk to each other about their work so they can recognize similarities and differences in their thinking.

As participants work with such diagrams, help them articulate the connections between the diagram and the statement by asking, "How does your diagram represent square numbers?" "How is multiplication indicated in your diagram?" Also ask questions that encourage participants to explore the generality of their arguments. "This diagram is built with the numbers 9 and 16. How does it show that *any* two square numbers will work this way?" They may argue that this process results in a single rectangle that has the same number of units on each dimension, thus assuring that the product is a square number.

Help participants articulate how the arguments satisfy the Criteria for Representation-Based Proof. Ask how multiplication is represented, how this representation applies to the general case, and how the arguments support the more general claim.

Proving using algebraic notation and the Laws of Arithmetic: Some small groups may have worked on a symbolic method to express their arguments. The statement begins with $(a^2) \times (b^2)$. That can be rewritten as $(a \times a) \times (b \times b)$, which can be rewritten as $(a \times b) \times (a \times b)$, or $(a \times b)^2$.

A formal argument explicitly calling upon the Laws of Arithmetic might take this form:

$(a^2) \times (b^2)$	Definition of a square number
$(a \times a) \times (b \times b)$	Definition of an exponent
$[(a \times a) \times b] \times b$	Associative Property
$(a) \times (a \times b) \times (b)$	Associative Property
$(a) \times (b \times a) \times (b)$	Commutative Property
$[(a \times b) \times a] \times b$	Associative Property
$(a \times b) \times (a \times b)$	Associative Property
$(a \times b)^2$	Definition of an exponent

(*Note:* In this argument, each step has been written out. Frequently, steps are combined and justified by citing that repeated use of the Associative and Commutative Properties guarantees that factors can be regrouped and reordered.)

Ask questions about the connections between the diagrams and the symbolic argument. Some participants may find that the symbolic argument, as well as the Representation-Based Proof, illustrates a stronger principle than the original statement. Not only is the product of two square numbers also a square number, but you now know *which* number is squared. The value of the resulting square number can also be determined, and its relationship to the original square numbers is evident. It is the square of the product of the square roots of the original square numbers; i.e., the product of 9×16 is $(3 \times 4)^2$, or 12^2. Once this point has been made, invite participants to look at the diagram solutions to see how this conclusion is captured in that representation as well.

Final Reflection Questions

This is an opportunity for you to think through your experiences in the seminar. Read through your collection of assignments, the facilitator's responses, your exit cards, and your math work to get a sense of how your ideas have changed. Use this body of work to guide your responses to the following questions:

1. How have your ideas about algebra changed over the course of the seminar? Be specific about the ideas that have changed.

2. How have your ideas about learning changed? Be specific.

3. How have your ideas about teaching changed? Be specific.

4. It is likely that there are issues—both mathematical and pedagogical—that came up during the seminar that continue to puzzle you. Pick one issue that is still "alive" for you. Explain what it is and your current thinking about it.

Evaluation Form

Please respond to the following questions, and return your responses together with your final reflection in the addressed and stamped envelope.

1. What did you like about the way the seminar was conducted? Be specific.

2. What aspects of the seminar did you not like?

3. If the seminar were offered again, what changes would you suggest to help make the experience more beneficial for participants?

4. What other thoughts would you like to share with seminar facilitators about your experience?